AI Innovations for Improving the Food Industry

Pethuru Raj
Reliance Jio Platforms Ltd., Bangalore, India

Kai Sheng
Xidian University, China

Kavita Saini
Galgotias University, India

S. Koteeswaran
S.A. Engineering College, India

IGI Global
Scientific Publishing
Publishing Tomorrow's Research Today

Vice President of Editorial — Melissa Wagner
Director of Acquisitions — Mikaela Felty
Director of Book Development — Jocelynn Hessler
Production Manager — Mike Brehm
Cover Design — Jose Rosado

Published in the United States of America by
IGI Global Scientific Publishing
701 East Chocolate Avenue
Hershey, PA, 17033, USA
Tel: 717-533-8845 | Fax: 717-533-7115
Website: https://www.igi-global.com E-mail: cust@igi-global.com

Library of Congress Cataloging-in-Publication Data

Names: Raj, Pethuru, editor. | Sheng, Kai, 1976- editor. | Saini, Kavita, 1976- editor. | Koteeswaran, S., 1982- editor.
Title: AI innovations for improving the food industry / edited by Pethuru Raj Chelliah, Kai Sheng, Kavita Saini, Koteeswaran S.
Description: Hershey, PA : IGI Global Scientific Publishing, [2026] | Includes bibliographical references and index. | Summary: "The food industry consciously leverages AI to be correct and relevant to customers and consumers. AI comes in handy and brings numerous advantages to food companies worldwide. Experts strongly believe that AI-powered solutions and services can guarantee a drastic surge in operational efficiency. Further, AI can conceive and concretize premium and people-centric services and explore fresh avenues for additional revenues"-- Provided by publisher.
Identifiers: LCCN 2025004047 (print) | LCCN 2025004048 (ebook) | ISBN 9798369398845 (h/c) | ISBN 9798369398852 (s/c) | ISBN 9798369398869 (ebook)
Subjects: LCSH: Food industry and trade--Automation | Food industry and trade--Data processing | Food industry and trade--Technological innovations | Artificial intelligence--Industrial applications | Artificial intelligence--Economic aspects
Classification: LCC TP372.8 .A35 2026 (print) | LCC TP372.8 (ebook)
LC record available at https://lccn.loc.gov/2025004047
LC ebook record available at https://lccn.loc.gov/2025004048

British Cataloguing in Publication Data
A Cataloguing in Publication record for this book is available from the British Library.

Table of Contents

Detailed Table of Contents

Artificial Intelligence is increasingly becoming a transformative force in the food industry, reshaping how food is produced, processed, and delivered. The current research on the impact of Artificial Intelligence on improving operational effectiveness in the food industry. Firstly, it emphasizes how AI technologies enhance production processes and optimize supply chain management while prioritizing food safety. By employing predictive analytics and automation, AI contributes to minimizing waste, maximizing resource use, and upholding high-quality standards throughout the production cycle. Secondly, the study investigates how AI boosts customer interaction by offering personalized experiences and streamlined service delivery. By evaluating consumer preferences and behaviors, AI allows food enterprises to customize their offerings, ultimately resulting in heightened customer satisfaction and loyalty. Through this investigation, the research seeks to provide practical insights and best practices for food sector stakeholders aiming to exploit AI for better operational results.

AI is gradually becoming the most powerful tool in the food and agriculture industries due to the possibility of making informed decisions on various processes in the supply chain. Machine learning, computer vision, and robotics are widely used in agriculture to increase productivity, optimize crop yields, food safety, and apply automatization in the process of food processing as well as provide individualized

food inspection, recognition, spoilage, agriculture, and nutrition recommendations. Disparate concepts of DL architectures are also discussed as well as the recent popular trends such as federated and multimodal learning and its integration with IoT and edge computing. The chapter also describes the issues concerning the quality of data, its interpretability, deployment, and the ethical issue. Finally, we provide insights on research themes that would help in the development of food systems that are sustainable, protective of the consumer's privacy, and efficient. This chapter explores how DL can be used to power innovation and resilience in the food system that is rapidly becoming data-oriented.

Improving the food industry requires innovative techniques for quality assessment, particularly in evaluating the characteristics of food products such as coffee beans. This study explores advanced methodologies for analyzing coffee roasting levels through image processing, feature selection, and classification. Coffee bean samples from three roasting levels—overdeveloped, underdeveloped, and standard—were imaged and preprocessed using Gaussian distribution for noise reduction and Laplacian filtering to enhance feature clarity. Feature extraction and selection were conducted using Particle Swarm Optimization (PSO) and the Butterfly Optimization Algorithm (BOA), ensuring the identification of relevant attributes while minimizing redundancy. Additionally, a Convolutional Neural Network (CNN) was employed for color analysis, providing a robust assessment of quality based on visual features.

This chapter aims to explore the transformative role of AI-powered drones in the food industry, particularly in the context of Industry 5.0. The primary objective is to examine how AI drones can optimize food production, enhance logistics, improve sustainability, and ensure food safety. The target audience includes researchers, food industry professionals, AI engineers, and business executives interested in

Preface

INTRODUCTION

In recent years, artificial intelligence (AI) has rapidly evolved from a trailblazing data analytics technology into a generative and increasingly creative force. With advancements in foundational models and tools such as generative adversarial networks (GANs), variational autoencoders (VAEs), and the widespread availability of specialized processing hardware—GPUs, TPUs, NPUs, and VPUs—AI is no longer a futuristic concept, but a practical and powerful technology transforming entire industries.

Across the business landscape, AI is being recognized as a strategic differentiator. It enables companies to refine operations, rethink services, and reimagine customer experiences. Coupled with other transformative technologies—5G, edge computing, IoT, robotics, blockchain, miniaturized sensors, and more—AI is driving sophisticated and scalable solutions that once belonged to science fiction.

Among the industries undergoing this transformation, the food industry stands out for its dynamic and multidimensional adoption of AI. From enhancing food safety and optimizing supply chains to enabling personalized nutrition and sustainable production practices, AI is playing an increasingly critical role. It brings the promise of higher efficiency, greater consumer relevance, and innovative offerings that were previously unattainable.

This edited volume, *AI Innovations for Improving the Food Industry*, seeks to document and explore this ongoing evolution. It presents the latest research, practical implementations, and forward-looking insights into how AI technologies are being applied to solve real-world challenges in the food sector. Through a curated set of chapters contributed by experts from academia and industry alike, we delve into both the theoretical underpinnings and practical implications of AI integration.

Our intended audience spans a broad spectrum:

- Researchers and academics exploring the applications of AI in industrial settings, particularly in addressing pressing issues like sustainability, food safety, and product innovation;
- Data scientists and AI engineers are developing machine learning, computer vision, and predictive analytics models tailored to the specific needs of the food ecosystem.
- Industry professionals and executives are aiming to leverage AI to streamline operations, improve decision-making, and enhance competitive advantage.

AI provides a powerful toolset for making informed decisions, empowering professionals to navigate the complexities of the food industry with confidence.

This book serves as both a reference and a roadmap for those seeking to understand and shape the future of the food industry through AI. We hope that the knowledge and perspectives shared within these pages will spark new ideas, foster cross-disciplinary collaboration, and ultimately contribute to a more innovative, efficient, and sustainable food system.

ORGANIZATION OF THE BOOK

Chapter 1: Benefits of AI in the Food Industry

This foundational chapter introduces the transformative impact of AI across food production, processing, and delivery. It explores how AI technologies like predictive analytics and automation enhance operational effectiveness, reduce waste, and uphold food safety. In addition to optimizing internal processes, the chapter delves into AI's role in enhancing customer experiences through personalization and behavior-based services. Practical insights and best practices are shared with industry stakeholders seeking to leverage AI for enhanced efficiency and customer loyalty, inspiring further research and exploration in this exciting field.

Chapter 2: Artificial Intelligence in Food and Agriculture: Driving Sustainability, Efficiency, and Innovation

Focusing on AI's applications in agriculture and food production, this chapter highlights the use of machine learning, robotics, and computer vision in areas such as crop yield optimization, food safety, and sustainable processing. It offers a comprehensive survey of key domains, current innovations, and ongoing challenges in AI adoption. Special attention is given to AI's potential in solving global food security issues and resource management challenges amid climate change, positioning AI as a pivotal tool in the quest for sustainable development.

Chapter 3: Machine Learning for a Smarter Food Industry: Trends, Challenges, and Innovations

This chapter presents a deep dive into the expanding role of machine learning (ML) in modern food systems. Covering a broad spectrum—from agricultural forecasting to personalized nutrition—it analyzes commonly used ML models, emerging trends like explainable AI and federated learning, and pressing concerns such as data privacy and scalability. By identifying both the opportunities and constraints of ML adoption, this chapter paves the way for future research directions aimed at creating intelligent and adaptive food systems.

Chapter 4: Harnessing Deep Learning for a Smarter and Sustainable Food Industry

Deep learning (DL) is playing an increasingly critical role in automating complex tasks across the food supply chain. This chapter examines the applications of DL in food inspection, spoilage detection, agricultural monitoring, and nutrition advising. It also addresses the growing integration of DL with IoT and edge computing, while confronting challenges related to data quality, interpretability, and ethics. The discussion emphasizes the potential of DL to foster sustainability and consumer trust in data-driven food ecosystems.

Chapter 5: Optimization Algorithms in Food Product Quality Assessment

Exploring innovative techniques in quality evaluation, this chapter presents a detailed case study on assessing coffee bean roast levels using image processing and metaheuristic optimization. Through methods like Particle Swarm Optimization and the Butterfly Optimization Algorithm, the study demonstrates how AI can improve accuracy and efficiency in food quality control. A CNN-based color analysis further enriches the methodology, illustrating the power of combining AI with sensory evaluation in food production.

Chapter 6: AI-Powered Drones for the Food Industry: Revolutionizing Food Production, Logistics, and Sustainability

Drones integrated with AI are emerging as a disruptive force in food production and logistics. This chapter examines real-world applications of AI-powered drones across precision agriculture, supply chain optimization, and environmental monitoring. Drawing from case studies and expert interviews, it offers evidence of improved efficiency and reduced environmental impact. However, it also addresses regulatory, technological, and infrastructural barriers, providing a balanced view of the opportunities and limitations.

Chapter 7: Plant Disease Detection for Food Safety

Plant diseases remain a critical threat to food security, and this chapter focuses on how AI—particularly machine learning—can transform disease diagnostics. It critiques conventional diagnostic methods and introduces data-driven techniques that improve speed and accuracy. By enabling early detection and timely intervention, AI tools empower farmers and researchers to mitigate crop loss and safeguard food supplies, making a compelling case for the broader deployment of AI in agricultural health monitoring.

Chapter 8: Smart Dining: AI-Powered Innovations Shaping Customer Experience in the Food Industry

Customer interaction is undergoing a digital renaissance, and this chapter explores the role of AI in shaping the next-generation dining experience. From intelligent chatbots and voice assistants to recommender systems and emotion recognition, AI enhances personalization and operational responsiveness. The chapter also addresses ethical considerations like data privacy and algorithmic bias, offering a holistic view of how AI can be employed to create smarter, more satisfying, and ethically sound dining environments.

Chapter 9: Integrating AI for Tea Leaf Classification: A Case Study in the Food Industry

This chapter presents a practical case study on using computer vision and deep learning to classify tea leaves by quality and type. With advanced preprocessing, feature extraction via Particle Swarm Optimization, and classification using ResNet-50, the study exemplifies AI's precision and efficiency in food inspection. It demonstrates how targeted applications can elevate quality assurance, reduce manual error, and drive operational improvements in specialized sectors of the food industry.

CONCLUSION

As the food industry navigates an era marked by rapid technological change, artificial intelligence stands at the forefront of its transformation. Through the chapters presented in this volume, it becomes abundantly clear that AI is no longer a distant possibility or an experimental add-on—it is now a practical, scalable, and essential tool for addressing the sector's most pressing challenges and unlocking its most significant opportunities.

From enhancing supply chain efficiency and quality assurance to driving personalized consumer experiences and promoting sustainable practices, AI's impact is both wide-reaching and profoundly transformative. Whether through machine learning, deep learning, generative models, or AI-integrated hardware like drones and sensors, these technologies are converging to redefine how food is produced, processed, distributed, and consumed.

However, this transformation does not come without questions. Ethical concerns, data privacy, model interpretability, infrastructure limitations, and regulatory compliance are just some of the issues that demand thoughtful consideration. As AI becomes further embedded in the food ecosystem, interdisciplinary collaboration among technologists, food scientists, policymakers, and industry leaders will be essential.

This book is intended as both a resource and a catalyst. It provides current knowledge, practical case studies, and forward-looking perspectives to support researchers, engineers, professionals, and decision-makers in harnessing AI's full potential. We hope it inspires innovation, informs strategy, and fosters dialogue across disciplines, ultimately contributing to a food system that is smarter, safer, more resilient, and more aligned with the needs of people and the planet.

We thank our contributors for their insights and dedication, and we look forward to the continued evolution of AI applications that will shape the future of the global food industry.

Chapter 1
Benefits of AI in the Food Industry

D. Hemanth
https://orcid.org/0009-0002-6630-3103
Christ University, India

ABSTRACT

Artificial Intelligence is increasingly becoming a transformative force in the food industry, reshaping how food is produced, processed, and delivered. The current research on the impact of Artificial Intelligence on improving operational effectiveness in the food industry. Firstly, it emphasizes how AI technologies enhance production processes and optimize supply chain management while prioritizing food safety. By employing predictive analytics and automation, AI contributes to minimizing waste, maximizing resource use, and upholding high-quality standards throughout the production cycle. Secondly, the study investigates how AI boosts customer interaction by offering personalized experiences and streamlined service delivery. By evaluating consumer preferences and behaviors, AI allows food enterprises to customize their offerings, ultimately resulting in heightened customer satisfaction and loyalty. Through this investigation, the research seeks to provide practical insights and best practices for food sector stakeholders aiming to exploit AI for better operational results.

DOI: 10.4018/979-8-3693-9884-5.ch001

INTRODUCTION

Artificial Intelligence (AI) is transforming the food sector in both exciting and significant manners. With the increasing global population, the demand for various types of food is rising rapidly. This presents challenges for the food industry, such as how to efficiently produce enough food, ensure its safety, and do so sustainably. AI technologies, which include machine learning, predictive analytics, and automation, are emerging to help address these issues and enhance the processes of food preparation, processing, and distribution can swiftly analyze extensive datasets, enabling businesses to make improved decisions and operate more efficiently. For example, AI can assist food manufacturers in scheduling their production more effectively, ensuring they produce the right quantity of food to meet demand while minimizing waste. This not only helps businesses cut costs but also lessens the environmental impact associated with overproduction.

One of the most significant advantages of AI in the food industry is its capacity to enhance supply chain management. Historically, it has been challenging for businesses to trace food products from the farm to consumers. However, AI can deliver real-time updates on the precise location of products at any moment. Consequently, if an issue arises, such as food contamination, businesses can promptly pinpoint the source and take necessary actions to protect consumers further enriching customer experiences by personalizing interactions. For instance, restaurants can utilize AI to analyze past orders and customer preferences, enabling them to suggest dishes that customers are likely to enjoy. Such personalization makes customers feel appreciated and can foster greater loyalty.

Nevertheless, many individuals in the food industry remain unaware of the potential benefits of AI. This lack of awareness can result in missed opportunities to enhance operations and improve decision-making. As a result, some businesses continue to encounter inefficiencies, food safety concerns, and a lack of personalized service. Envision a scenario where food producers can accurately forecast consumer preferences, allowing them to minimize waste and guarantee that fresh products are consistently available. Unfortunately, without a solid understanding of how to implement AI, such prospects may remain elusive.

Additionally, numerous businesses are entrenched in traditional practices, which can stifle innovation. An apprehension of change or uncertainty about new technologies can hinder companies from adopting AI solutions that could significantly benefit them. The potential for enhancement is substantial. By acknowledging the challenges and adopting AI, the food industry can become more efficient, safer, and more attuned to customer preferences.

Moreover, AI can advocate for sustainable methods by optimizing resource usage such as water and energy. For example, AI-powered smart farming techniques can assist farmers in monitoring their crops and soil, allowing for more effective use of fertilizers and pesticides. This not only increases crop yields but also safeguards the environment. In food processing, AI can automate repetitive tasks, enabling workers to concentrate on more creative and complex activities. This can lead to improved food quality and more innovative products that cater to evolving consumer tastes.

As we investigate the various ways AI can enhance the food industry, it becomes evident that embracing these technologies can facilitate a more efficient and resilient food system. By collaborating and utilizing AI, the food sector can not only meet the escalating demand for food but also build a better future for all stakeholders involved.

ARTIFICIAL INTELLIGENCE IN FOOD WASTE MANAGEMENT

Food waste is one of the global issues with major economic, social, and environmental implications. It is estimated that one-third of all food produced for human consumption is lost or wasted, which translates to around 1.3 billion tons annually. The waste does not only reflect the loss of valuable resources but also contributes to greenhouse gas emissions, thus worsening climate change. This challenge brought about the popularization of the circular economy idea, promoting more sustainable practices where waste is not created, or if it happens, then at maximum efficiency. There is now also a transformative power in the world of Artificial Intelligence (AI) that will deliver innovative solutions regarding food waste to improve the state of the circular economy. Thus, this document is an integral overview of how AI plays into food waste management, its application, challenges, and future aspects.

Understanding Food Waste

Food waste is a critical global issue with significant economic, social, and environmental implications. It refers to the discarding or loss of food that is fit for consumption, occurring at various stages of the food supply chain, from production to consumption. Understanding food waste involves examining its causes, impacts, and potential solutions to mitigate this pressing problem. Food waste is the total amount of food that is lost, wasted, or uneaten. It differs from food loss, which takes place at the production and supply chain levels. The reasons behind food waste are varied, such as production inefficiencies, like poor harvesting methods and inadequate storage facilities, as well as processing and distribution challenges, including spoilage and logistics. Retail practices also play a major role, since supermarkets throw away food that is close to expiring or is not aesthetically pleasing. On the consumer side, over-buying, failure to plan meals, and misconceptions about expiration dates are all driven by cultural preferences for fruits and vegetables that look perfect. The implications of food loss are enormous as it contributes about $936 billion in annual lost economic opportunities and supports greenhouse gas generation through decomposing food in the landfills besides wasting resources of water, lands, and energies used in growing food. Addressing food waste is important not only for improving food security and reducing hunger but also for promoting sustainable practices that benefit the planet and society as a whole.

The Role of Artificial Intelligence in Food Waste Management

Long before food waste became a worldwide problem that would represent approximately one-third of all the food produced worldwide for human consumption, it had always been perceived as being artificially intelligent. Presently, the emergence of AI is becoming one of the more revolutionary solutions over such a huge waste level that, only by using numbers, may be described as approximately one-third of all the food produced. AI improves food waste management throughout the entire food supply chain, from production to consumption, by using data-driven insights and automation.

Figure 1. Household food waste by region

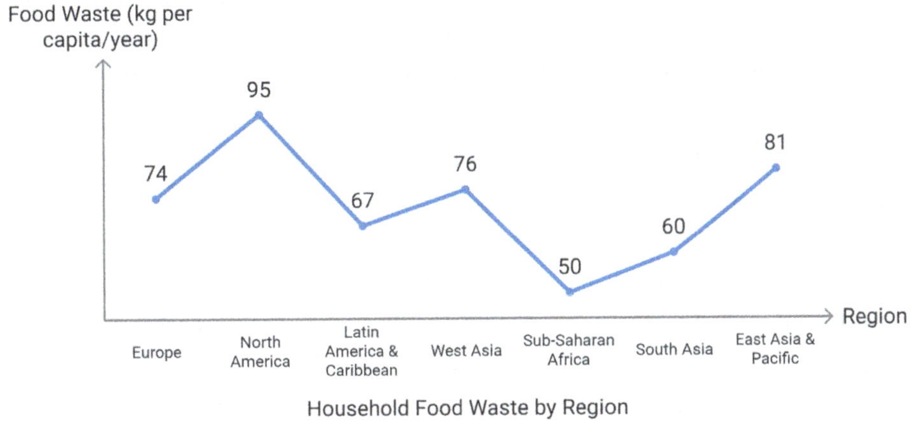

Household Food Waste Across Regions: Comparison

Regional differences in trends in household food waste are notable considering differences in consumer behavior, the capacity for handling waste, and culture. Indications in the data depicted in Figure X present the contrasts between the generation of food waste as depicted in kilogram per capita per year.

- North America has the highest per capita domestic food wastage of 95 kg per year, with excessive levels of overuse and dumping. Excessive food availability coupled with consumer attitudes amounts to excessive wastage.
- Europe has 74 kg per capita per year, although lower than North America, still represents a high wastage profile. Compulsory wast-

age reduction measures in most European countries, however, counterbalance excessive losses.

- Latin America & the Caribbean produce 67 kg per capita per annum of domestic food waste, marginally below Europe. Domestic food waste within the region is affected by economic inequities and variability in food distribution.
- West Asia produces 76 kg per capita per annum of domestic food waste, with seasonal fluctuations due to religious and festive periods leading to short-term increases in food disposal.
- Sub-Saharan Africa has the lowest per capita domestic food loss, at 50 kg annually that is due to food shortage, effective resource utilization, and use of conventional waste management strategies like animal feed and composting.
- South Asia loses 60 kg per capita annually, reflecting a comparatively modest rate of food wastage. There is inefficient food storage and distribution structurally that leads to wastage even with cultural inclinations towards being efficient.
- East Asia & Pacific experiences an intense increase, to 81 kg per capita annually. Japan, for example, has experience reducing food waste through policy responses, but rapid metropolitanization on a high level and shifting eating patterns are yielding huge amounts of domestic waste throughout Monitoring and Optimizing Food Production

AI plays a crucial role in monitoring and optimizing food production processes, significantly reducing waste at the source. Precision agriculture uses AI technologies, like machine learning, computer vision, and data analytics, to accumulate the data itself. The data is collected from multiple sources: soil sensors, drones, and satellite imagery. The data encompasses crop health, soil, and environmental factors, making farmers better decision-makers. For instance, AI algorithms can monitor the amount of moisture in the soil and weather predictions to schedule proper irrigation to provide crops with sufficient water without producing excess runoff. Another is that nutrient deficiency or pest infestation is identified at the earliest so intervention is done promptly, hence preventing the crops from being lost. In the optimization of resources as well as productivity, AI avoids waste in the agricultural production line.

With precision agriculture that integrates AI, the yield of the crops improves, and further enhances aid in the practice of sustainable farming. The application of fertilizers and pesticides in the areas they are required will reduce chemical runoff to waterways, thereby minimizing environmental pollution. In addition, AI systems can analyze history data to predict future crop performance and allow for farmers to adapt planting strategies based on such forecasts. This way of farming promotes efficiency in the use of resources and, thus leads to the availability of less quantity of food waste.

Global Progress in Food Waste Reduction

Figure 2. Global progress in foot waste reduction

- **2007:** The UK launched its first national initiative.
- **2008:** Japan introduced similar strategies.
- **2020:** Japan cut food waste by **31%** through redistribution and supply chain improvements.
- **2021:** The UK reduced waste by **18%** via packaging, portion control, and retail initiatives.

Redistribution of Surplus Food

The other role that AI plays in the management of food waste is in ensuring the redistribution of surplus food. In most cases, surplus is created at various levels of the supply chain-the point of production, the retail sector, among others. AI identifies this surplus then contacts relevant organizations to distribute this food to those in need. AI algorithms examine data concerning the availability, demand, and networks of distribution to find chances to re-route food supplies. For example, AI could determine the surplus food stores on hand and link these to a local food bank or other similar non-profit service organizations feeding needy populations.

AI can help improve logistics about food redistribution based on examination of routes, delivery schedule, and storage facilities. This helps minimize spoilage and waste because AI ensures the surpluses reach their destinations fast and in good time. This is especially true for perishable products, which need prompt distribution to stay within the time frame for consumption. AI can facilitate the redistribution of foodstuffs and reduce waste while meeting food insecurity in many communities, as most struggle to have access to healthy food. The AI-driven redistributed foodstuffs can, therefore, have a high impact on meeting food security.

AI is important in the food industry context in terms of circular economy initiatives that lack the generation of wastes and maximize resource use. Circular economies are promoted in practices like recycling and reusing, reducing waste, and AI technologies can study data to identify possible efficiencies of the use of food supply chain resources and their consumption. For instance, AI can help food processors analyze their process to see how they can minimize amounts of waste generated. AI-based solutions make sustainability better and the footprint smaller.

Recycling processes are upgraded as well. AI scans the incoming waste streams to segregate the materials that can be recycled and recovered for further reutilization. AI algorithms differentiate between food wastes that are categorized either for composting, recycling, or repurposing. It reduces the fill of landfills while preserving resources recovered. The use of AI enhances the circular economy by promoting business strategies that help firms to practice green economies and their impact on nature and the economy.

The integration of AI into circular economy practices also encourages innovation in product design and packaging. Based on the analysis of consumer preferences and market trends, AI can help companies design products that are more sustainable and less likely to contribute to waste. This approach to product development is proactive, ensuring that resources are used efficiently and waste is minimized, aligning with the principles of the circular economy.

Efficient Supply Chain

AI significantly improves the efficiency of the supply chain, which is very important in reducing food waste. The food supply chain is complex, involving many stakeholders and processes, and inefficiencies can lead to significant waste. AI-driven demand forecasting allows food companies to predict consumer demand accurately by analyzing historical sales data, market trends, and external factors. This enables companies to adjust their production and inventory levels accordingly, reducing overproduction and spoilage.

Additionally, AI optimizes delivery routes and schedules to ensure that perishable products reach consumers in the shortest time possible. AI analyzes traffic patterns, weather conditions, and delivery constraints to suggest the most efficient routes, thereby minimizing delays and reducing carbon emissions. This optimization improves the efficiency of the supply chain while also making it more sustainable by reducing the environmental impact of food distribution.

Potential Applications of AI in Food Waste Management

AI is revolutionizing many sectors, and its implementation in the food waste management sector also appears to be very promising. The new technologies can be incorporated into the chain of food supply to enhance the optimization of processes, increase efficiency, and therefore decrease the level of food waste.

Figure 3. Global food waste distribution in 2024

Global Food Waste Distribution in 2024: An Analysis

- The 2024 food wastage distribution, as presented in Figure 3, emphasizes the three primary world sources of food loss: households, food service establishments, and food retail stores. The data provides a pivotal understanding of food wastage creation and where intervention will be most effective.
- Domestic Waste (60%): The consumer level is the largest source of wastage of food with the highest percentage being from rotten food because it has become rotten, people buying food in excess, and because people have no idea about preserving food. This would be consumers' education, better portioning, and better ways of storing food.
- Food Service Waste (28%): Food service like restaurants, hotels, and caterers create nearly a third of global food wastage. Factors include over-preparation, loss through buffet style meals, and consumer leftover plates. Controls for portioning, donation schemes, and improved kitchen stock control are solutions to reduce wastage.

- Retail Waste (12%): Although retailers contribute the least to overall food waste, this 12% is still a lot. Causes include food spoilage before sale, strict aesthetic standards for fruits and vegetables, and overstocking perishables. Dynamic pricing, food bank redistribution of surplus food, and AI-driven inventory management can reduce retail-level wastage.

Machine Learning for Predicting Food Demand

ML is one of the most important subsets of AI, which helps in developing algorithms for learning and predicting from data. This aspect in food waste management is applied to predict the demand for food, thus leading to optimal levels of inventory with lesser overproduction.

Data Analysis and Forecasting. Here, the machine learning algorithms will analyze millions of historical sales data, trends, seasonal variation, and others like economic indicators and behavior of consumers. The ML model then arrives at much more precise future food demand forecasts by picking up all the patterns. For instance, in holidays or special events, ML can help the grocery shop predict how much of that particular product it will need in advance so that they can stock up in due time.

Advantages of demand forecasting for producers and retailers; they produce a lesser quantity to avoid much unnecessary waste if they successfully and accurately forecast what there is for sales. Perishable commodities go bad not in the course of sale when too much food in stock but otherwise. There could also be proper operational performance which saves extra spending on getting rid of leftovers in excess with less waste products due to inappropriate manufacture.

There are many examples of companies using ML for demand forecasting successfully. For example, large supermarket chains use ML algorithms to analyze the buying patterns of their customers and update their stock in real-time. This not only reduces food wastage but also makes sure that popular items are always available to customers.

AI-Based Smart Agriculture Techniques

AI-driven smart agriculture techniques involve the latest technology for optimizing farm practices, crop yields, and minimizing waste. These techniques apply precision farming, data analytics, and automation to a wide range of applications. Precision Farming Precision farming leverages AI technologies to monitor and manage the heterogeneity of the field for crops. Sensors, drones, and satellite imagery collect data regarding soil conditions, crop health, and other environmental conditions in real-time. These data feed AI algorithms that understand and provide actionable insights on irrigation, fertilization, and pest control decisions to the farmer.

AI-powered systems can optimize resource usage as they determine the exact amount of water, fertilizers, and pesticides needed for each crop. For example, AI can read soil moisture levels and weather forecasts to better schedule irrigation to ensure that crops get the right amount of water without excess runoff, which saves water and minimizes over-fertilization that may cause nutrient runoff and pollution in the environment. Techniques in smart agriculture make farmers more productive as well as reduce food waste at the production level. Since there is close monitoring of crops and resource application, overproduction followed by probable spoilage becomes less likely. AI can help farmers detect early problems such as infestation by pests or diseases, which definitely decreases crop loss to a great extent.

Natural Language Processing for Customer Review Analysis

NLP is a branch of AI that deals with the interaction between computers and human language. In the domain of food waste management, NLP can be applied to analyze feedback from customers and reviews made by them to know more about consumer preference and their behavior. Analyzing consumer sentiment: The NLP algorithm can process volumes of text-based data from diverse sources, which include social media, online reviews, and surveys of customers. By analyzing this information, businesses are able to understand what consumers feel about food products.

For example, NLP identifies common themes from customer feedback: freshness, quality, or packaging.

Knowledge regarding the consumer need will improve food companies' understanding of the consumer's expectation regarding their product. For instance, if a food commodity constantly complained about because it is not fresh as required, then there must be proactive work by that company to rectify the same issue with the help of an improved supply chain or perhaps rescheduling the manufacturing schedule. This improves satisfaction and waste from unsold or unpalatable products. Now, many food companies are using NLP to analyze consumer feedback. Restaurants and food service providers can monitor reviews online and mentions on social media through NLP, which will enable them to identify areas that need improvement and adjust their menu according to the change. This will help reduce food waste by offering the right product according to consumer demand.

IMPROVING QUALITY CONTROL

It would constitute improving the quality of food production's prime foundation: manufacturing food. This assures that food product prepared passes by safety perception through consumers. As recent times can illustrate, there was a food manufacturing company utilizing significant technological devices adequately equipped with AI and advances data analytics with regards to food product quality-control processing. This chapter addresses two primary applications of AI: automated quality inspection and data-driven quality management. This list of applications helps us realize how AI makes an impact on the quality assurance of the food sector and enhances the consistency of products being produced, reduces wastes, and increases consumer confidence.

Automated Quality Inspection

Automated quality control has changed the grading of a food product into an improved quality and safety. Compared to traditional mechanisms of quality control that involve human inspections, which are quite time-consuming and labor-intensive, prone to human faults, computer vision systems based on AI may offer approaches that are much more efficient and accurate in inspection of quality.

Implementation of Computer Vision Systems for Automated Quality Control

Computer vision technology applies AI algorithms for processing images and videos. Thus, computers can view images or video frames like human eyes and identify what is in the contents. In food processing, computer vision refers to the application of the system to inspect products along a production line for defects, inconsistency, or contamination. It is usually paired with high-resolution cameras and more complex image processing algorithms to enable them to spot slight anomalies against the standard of quality.

For instance, in a fruit processing plant, computer vision might be used for inspection purposes on how the fruits arriving on the conveyor belts look. A system will indicate bruising and discolorations or other defect areas that are considered unsoldable. Therefore, this, automatically cuts out the time, as well as the labor needed to check qualities, and raise the accuracy regarding defects detection at a much improved level (Kumar, Singh, & Kumar, 2020).

Benefits of AI in Detecting Defects and Ensuring Product Consistency

Some of the advantages of AI-powered automated quality inspection systems include the following: first, these systems speed up the process of quality control by several folds. Computer vision systems can scan thousands of products per minute and identify defective items at once to call for immediate action. Such rapid inspection capabilities are very useful in high-volume production environments where product quality needs to be sustained (Ding et al., 2023).

This implies the quality of detection improves through AI systems. This is because human beings tend to miss minor defects due to tiredness or lack of attention; however, algorithms analyze visual data accurately, thereby having a huge decrease in false positives and negatives-essentially, it ensures that only products that meet the quality standard will reach consumers. For instance, the bakery will scan using computer vision baked goods, such as it seeks only items perfectly shaped and colored, which are packed. That way, quality is assured with the final output (Zhang, Wang, & Zhang, 2021). The quality inspection automation systems are likely to yield vital data to enhance improvement in processes. On analyzing inspection data collected, there are likely recurrent defects and patterns to help them take the best production decisions for these processes. These allow for making proactive modifications into practices related to the manufacture of such products and achieve the desired levels of product homogeneity with least wastes (Kumar, Singh, & Kumar, 2020).

Data-Driven Quality Management

In addition to automated inspection, AI is significantly contributing to data-driven quality management. Food manufacturers will be able to understand their production processes by using AI algorithms that analyze quality data and improve the production process to enhance overall quality assurance. Data-driven quality management refers to the organized gathering and analysis of quality data in production, for all points to be extracted, as well as all the interlinked relationships, to be known. The huge amount of data collected from logs on production, results of inspections, and environmental conditions can easily be processed through the AI algorithm and applied with the appropriate machine learning methods to extract complex patterns and relationships, some of which cannot be inferred by other types of analysis alone (Ding et al., 2023).

For example, a dairy production process would have milk quality information as; temperature, pH levels, and microbe counts. Using such algorithms, artificial intelligence will discover the ideal condition of milk production quality throughout all the steps within the process. As a result, the changes will meet with quality standards when sustained in due course (Zhang, Wang, & Zhang, 2021).

INCREASE EFFICIENCY IN PRODUCTION AND PROCESSING

Process improvements related to efficiency for production and food processing have also been in place, in support of growing consumers' demands to avoid increased cost and environmental problems. Artificial intelligence is the future in this business; it equips food manufacturers in improving operations efficiency, reducing loss, and becoming productive. Therefore, this article covers AI contribution to process improvements that have impacts of less energy with the elimination or reduction of the wastage phenomenon.

Process Optimization

The scope of operations in food processing changes through the insights and automation provided by AI to improve efficiency. Amounting to enormous data produced from production, AI technologies utilize this trend and patterns spotting ability to advise on improvement opportunities in decision-making. Food manufacturers can now efficiently upgrade their overall operation, improving ingredient usage, process times, and equipment performance using AI.

Predictive analytics is one of the significant tools through which AI offers value to process optimization. For example, using historical data, an AI algorithm can predict future production needs, and thus, manufacturers can adjust their processes. For instance, AI can predict seasonal trends of fluctuations in demand for certain products. Therefore, companies can maximize production schedules and levels of inventory. This proactive approach prevents overproduction, hence saving a lot of wastage, which makes the operation effective and efficient (Ding et al., 2023).

Examples of Use of AI for Energy and Waste Saving

Energy Management Systems: Artificial intelligence energy management systems are also used in food processing locations to monitor and optimize energy usage. It evaluates the data in real time based on how a machine or tools read it. It detects flaws and suggests corrections to the system to rectify such mistakes. For example, AI can run refrigeration machines at peak performance by regulating their temperatures based on current demand, significantly saving energy. Zhang, Wang, and Zhang (2021) showed that AI-based energy management systems in a dairy processing plant reduced energy costs by 15% by optimizing cooling processes. It simply speaks to smart manufacturing by making interlinked production systems on the basis of AI and IoT technologies. They are controlling production lines of foods based on input information from all kinds of data from machinery as well as different sensor-related inputs to their supply chains. For instance, baking process in a bakery can be controlled and oven temperature along with baking time could be altered by the real-time data so that energy consumption becomes minimal for quality products. Thus, not only is the quality ensured but also lesser wastage through overbaking or underbaking is guaranteed (Kumar, Singh, & Kumar, 2020).

AI significantly helps in reducing waste through predictive maintenance. Predictive maintenance of equipment is about the algorithms applied by AI when analyzing data from machinery to predict when the equipment will likely break down or when it will require maintenance. It would, in turn, help manufacturers to plan at non-peak hours for the maintenance schedule and, thus reduce downtime and the loss of production. For example, a meat processing plant would avoid equipment failure that leads to product spoilage and, in the process, reduces waste as well as saves cost in the general run (Zhang, Wang, & Zhang, 2021).

Ingredient Usage Optimization

AI technologies ensure that the amount of ingredients needed in food manufacturing is optimized thus reducing waste as well as cutting costs. For example, AI algorithms can analyze recipes and production processes to determine the best quantities of each ingredient needed in a batch. The manufacturers can cut down on the waste and thus increase profitability. For instance, a beverage manufacturer example showed that AI-powered ingredient usage optimization can reduce 10% of raw materials for similar quality of products (Ding et al., 2023).3.1 Applications of AI in Energy Intake Minimization and Waste Generation

Applications of AI in energy intake minimization and waste generation in food production and processing. Examples include:

Energy Management Systems

AI-based energy management systems can track and optimize the consumption of energy in food processing facilities. Such systems monitor real-time data from machinery and equipment to look for inefficiencies and recommend fixes. For example, a dairy processing plant may use AI to optimize refrigeration units by varying temperatures based on real-time demand, thereby resulting in significant saving in energy usage. Zhang, Wang, and Zhang (2021) demonstrated that AI-based energy management systems lowered energy costs by 15%.

Smart Manufacturing: AI and IoT are used in smart manufacturing to monitor and optimize the process in real time. For example, AI may identify patterns from data generated on production lines, such as the energy waste because of the running of equipment at suboptimal levels. According to this analysis, manufacturers can decrease energy consumption without affecting the quality of the product by adjusting operational parameters. A case study of a snack food manufacturer demonstrated that AI-driven process optimization led to a 20% reduction in energy usage during production (Kumar, Singh, & Kumar, 2020).

The ability to analyze historical production data using predictive analytics means AI can develop trends and predict what is needed going forward, hence optimizing the usage of ingredients in a particular product and reducing waste. For instance, AI can be used by a beverage company to accurately calculate demand for individual flavors at certain times of the year because of seasonal demand, thus ensuring the production corresponds to consumer preference. This will prevent excess production and also eliminate the associated waste from those items that will not be sold. According to a research paper, it has been stated that AI-based demand forecasting in a beverage firm reduced raw materials wastage by 10% (Ding et al., 2023).

Automated Quality Control

AI computer vision can be used to automatically inspect products so that only products of quality standard are packaged and shipped. Defects are picked at the production line, hence waste from rework and spoilage is significantly reduced. An example of an AI-powered inspection at a fruit processing facility led to a reduction of 30% of the defective products arriving in the marketplace, thus huge savings in terms of cost and reduction in waste (Zhang, Wang, & Zhang, 2021).

Predictive Maintenance

Predictive maintenance uses AI in forecasting failures within the equipment, and these take place proactively. Data coming from machines will be interpreted through AI algorithms so that certain patterns and anomalies will be spotted before an actual failure occurs. Sensors installed in a production device could collect information such as vibrations and temperature during operational runs. Then AI algorithms are used to scan the same data to provide alert of first signs of wear, thus scheduled maintenance period would occur before any kind of equipment failure.

This predictive ability prevents food manufacturers from facing sudden downtime, so they do not experience sudden stoppages. For example, a meat processing plant which began using predictive maintenance reduced its equipment failures by 25%. The operational efficiency increased due to the reduction of failures (Kumar, Singh, & Kumar, 2020).

Advantages in the Reduction of Downtime and Maintenance Cost

There is a significant amount of benefit that is derived from predictive maintenance. It allows for the pre-scheduling of maintenance to equipment so that it is possible to predict its probable failure, thus significantly reducing the downtime in food manufacturing operations. Therefore, it would also ensure that the production process was smooth and free from any form of product spoilage due to failure. Furthermore, predictive maintenance saves a lot of money. Manufacturers will not incur expensive repairs and replacements since the maintenance needs are addressed before they become significant problems. In one case study with a bakery, predictive maintenance resulted in a decrease of 30% in the maintenance costs, and there was better resource usage (Ding et al., 2023).

ECONOMIC BENEFITS OF AI ADOPTION IN THE FOOD INDUSTRY

Artificial Intelligence in the food industry changes the operational process and provides key economic benefits. The increasing demand with the problem of food safety and sustainability as critical concerns for the food industry, AI uniquely offers a solution that makes for efficiency, quality, and satisfaction of the customers. The following chapter investigates the cost savings AI can make on the operations within the food industry and thus represents a general financial review of AI adoption.

Cost Cutting

Labor Cost Savings

The first benefit in terms of economics that AI offers the food industry is saving on labor costs. Technologies in AI, such as automation, allow for streamlined processes that otherwise would involve human intervention. For example, robots can sort, pack, and check quality, using AI in place. As such, the company releases its human resources to more complex and value-added activities. This reduces labor costs while upping productivity in general. Labor costs can be one of the most significant portions of the operational expenses in traditional food production environments. Companies can save a lot by introducing AI-driven automation. For example, automated systems can work for extended periods without breaks, increasing output and efficiency. Additionally, the reduction in manual labor minimizes the risks associated with human error, thereby improving productivity and quality.

Efficiency

AI systems further optimize production operations by analyzing all data in real-time and providing adjustments to minimize inefficiencies to improve efficiency in production. Optimizing cooking time, temperature, and ingredient consumption can reduce a lot of wasted energy. Improving efficiency not only reduces operation costs but the quality of produced food is sustained. The AI can analyze huge volumes of data and make informed decisions. In fact, these AI algorithms could monitor production lines and identify bottlenecks or inefficiencies real-time. Quickly addressed, such bottlenecks may be, ensuring smooth operations so that the amount of downtime reduced. In such a case, efficiency not only reduces costs but also improves quality in the product being produced.

Predictive Maintenance

Another crucial economic benefit of AI in the food industry is predictive maintenance. Predictive maintenance based on AI avoids equipment failure since it predicts when maintenance is necessary to avoid breaking down the equipment. This preventive approach minimizes unplanned downtime and its accompanying emergency repair expenses. Maintaining equipment in the best possible state for their life cycle ensures cost savings for extended machinery life and lower maintenance expenses. Predictive maintenance applies data from machineries and equipment to learn patterns and, hence, determine when a part of the machinery is likely to fail. A piece of AI can analyze a history of a machine's performance and will determine when exactly that machine might need maintenance, allowing companies to schedule maintenance operations during off-peak hours for minimal disruption of production. Thereby, efficiency increases with reduced maintenance costs as well as minimizing the risk of costly breakdowns.

Supply Chain Optimization

AI makes the supply chain efficient in terms of demand forecasting and inventory management. Correct demand forecasting would avoid overproductions, hence no wastage and cost saving in that. The optimized inventory will reduce holding costs as well as risks of spoilage of perishables. The AI algorithms can analyze historical sales data, market trends, and weather patterns to establish a very appropriate demand forecasting. This information helps the companies to time their production schedule and inventory within the companies. Businesses minimize excess inventory and waste by synchronizing their production schedule according to actual demand, and this consequently leads to a considerable reduction in cost.

AI can also optimize logistics and distribution processes. AI systems analyze data related to transportation routes, delivery schedules, and customer demand to identify the most efficient ways to move products from manufacturers to retailers. Optimization reduces transportation costs and improves overall supply chain efficiency.

Energy Savings

Another vital expenditure in a food processing plant includes the aspect of energy, which AI-aided management systems monitor, optimize, and indeed bring down to enormous sizes. They track inefficiencies on real-time basis and recommend tweaks to energy intake. For example, the system can regulate how the refrigerating units operate with the refrigerator switched on only in times when usage is required with the intention to reduce the expense of energy related to cooling purposes. Through use of data gained from temperature and humidity, AI will ensure all refrigeration units are working, and therefore saving the energy lost. In the same way, AI will pinpoint other more energetic sectors such as cooking and heating processes, and these can too save more energy. It reduces consumption of energy while saving costs added to the overall sustainability efforts. As a result of reduced energy consumptions, food manufacturers decrease their carbon footprints and give in to demand from consumers constantly increasing for environmentally friendlier practices in the production process

PROMOTE SUSTAINABILITY

The food industry has reached such a stage in which sustainable practice is in higher demand than ever. Issues of climate change, resource depletion, and the rising demand for food before mankind have created a sense of urgency and a necessity to infuse sustainability into agricultural practice. Artificial Intelligence is perceived as the most transformative force that will be capable of generating sustainability in the food sector. AI technologies may aid in resource usage optimization, reducing waste and making it efficient.

Industry Reporting of Food Waste

Effective waste reduction efforts are reliant upon good reporting and measurement of food waste. Specific reporting processes are required in certain industries for the provision of credible data collection and analysis.

Household Methods

- "Surveys, waste composition surveys, and direct measurement methods" are best suited to gather household wastes information.
- "These methods allow trends of food wastage at the point of consumption to be established, with more precise reduction campaigns."

Food Service Practices

- Volumetric measurement and copycounting/scanning machines are best for food service establishments (restaurants, cafeterias, catering companies).
- The practices help in tracking portion size, spoilage, and excess food to aid in inventory control and minimizing waste.

Retail Practices

- Supermarkets and retail stores track waste using average mass balance, waste audit, and direct measurement.

- Methods assist companies in managing packaging waste such as waste, overstock, and expired stock.
- Direct measurement and waste monitoring equipment is mandatory for food processing and manufacturing industries.
- Procedures help ensure production waste is minimized, raw materials are used appropriately, and byproducts are treated accordingly.

Figure 4. Which sector's food waste reporting method should be implemented?

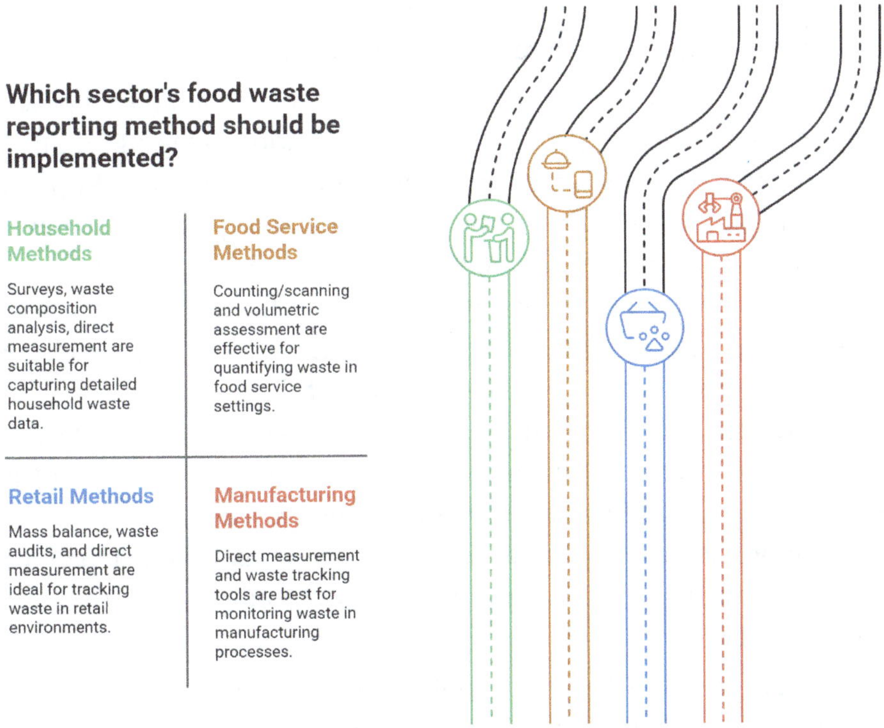

Sustainable Farming Practices

Sustainable Farming Practices Sustainability in farming is important to ensure that agricultural production will be able to meet the food needs at hand and in the future without causing harm to the environment. Most are cautiously optimistic about the major application of AI when concerning resource optimization and sustainability. AI technology improves the decision-making capacity of farmers based on real-time data analysis. From data accumulated through the soil sensors, weather forecasts and satellite imagery, AI derives apt suggestions on planting times, irrigation to crops, as well as crop health. Therefore, this gives rise to a data-based approach through which farmers can make optimal use of resources and thus reduce wastage with increased productivity.

AI-based systems monitor constantly soil condition and crop health. Examples include drones carrying high-resolution imaging equipment and integrating AI, which may be supplied to farm fields. Crop health assessment and areas requiring improvement can be possibly achieved in this manner. In this regard, it becomes possible to intervene in a timely manner by preventing and acting using measures like targeted watering or attacks by pests in their proper usage optimization. Such a system through AI-driven watering systems will have the capability of determining levels of soil water and weather forecast and thus provide crop irrigation with amounts that are correct, using water, when necessary, only so that it makes sure there's a reduction in water usage by irrigation. In research, it has been proven to be accurate that AI-based irrigation management is capable of reducing water usage for agriculture purposes entirely or by using it from as low as 30%.

AI can also help in fertilizer application through the analysis of nutrient levels in soil and crop needs. AI reduces the risks associated with over-fertilization that often causes the runoff of nutrients and leads to environmental pollution because it gives a specific recommendation for the application of fertilizer. Experiments in corn farms show that the levels of fertilizers applied can be reduced to as low as 20% of the usual need, and crops can still be maintained at the standard level. A significant advantage of AI precision agriculture is that water-saving benefits are associated with fertilizer-saving benefits. Real improvement in both water-use schedule and quality through the farmer where he does not waste any of the water resources, while delivering the crop exactly what is needed-the exact amount of moisture under poor water conditions.

The AI technologies used are sustainable and friendly to environmental stewardship. For example, for a farmer, optimizing resources will decrease the ecological footprint hence conserving the natural resources for future use. This is increasingly important to consumers as well as an interesting regulatory body, as a call for a greener form of food production is building up every day. Therefore, to put it in a nutshell, AI applications of precision agriculture give a new lease of life to sustainable farmlands as they make optimal usage while minimizing both water and fertilizers used to produce crops. Advanced monitoring predictive analytics and targeting resource management for better operations facilitate sustainability for agriculture. The cost-effectiveness of AI in water and fertilizer usage reduction benefits crop quality and conserves the environment. AI technologies will be the drivers of sustainable practices and solutions to the challenges of food production in a changing world.

Circular Economy Initiatives

The term "circular economy" has become a buzzword for the food industry-a more sustainable alternative of traditional linear mode of production and consumption. Such an economy will not consume resources efficiently; it produces less waste. Also, by making products as part of such designs, resources may be retrieved back through processes such as recycling, reuse, and composting. It is with such importance that AI has greatly contributed to the growth of the circular economy through the proper management of resources as well as reduction of wastes for value generation from by-products. The following section shall discuss the role of AI in supporting circular economy in the food industry, as well as some examples of companies using AI to make its implementation to sustainable practices.

The Role of AI in Support Circular Economy Practice in the Food Industry

Resource Optimization. All information sources regarding the food supply chain in general are factored into use by AI technologies. AI provides insights that help companies know whether their processes of production are inefficient and which ones to correct. For example, if AI can see patterns in energy consumption at food processing plants, companies can take measures to save energy, which will make it cheaper but also less harmful to the environment.

Waste valorization implies the conversion of waste material to a value product. For example, how waste food can be analyzed toward its byproducts could be used in organic waste, such as bienergy or even composted to make compost. Waste valorization shall reduce the volumes and provide enterprises involved with a fresh source of revenue. AI algorithms could decide which one of these routes of conversion of food waste into animal feeds or bioplastics would yield the highest outputs.

Supply Chain Transparency. AI enhances supply chain transparency by providing the latest available information regarding how products are being moved and where resources are being used. This will thus enable firms to trace the product life cycle throughout production all the way through to consumption. Opportunities to recycle and provide new lives to materials are therefore identified, and better knowledge of resource flows will guide such choices in accord with circular economic objectives.

Predictive Analytics in Demand Management. AI-based predictive analytics can help the company to take better decisions regarding demand management and thus avoid overproduction and associated wastage. The reason is that AI can make more accurate predictions of the consumer's demand with the help of sales history and trends in the market; thus, organizations can adjust the schedule of production in order to get aligned with actual demand, thereby cutting down unnecessary inventories and waste of food products, which also helps to attain circular economic objectives.

Some Examples of Organizations Using AI to Implement Sustainable Initiatives

Coca-Cola. It follows the circular economy with AI-based solutions in its supply chain management system without wasting the material. It utilizes AI to optimize routes for distributing the product and saves the emission cost on transport. The company has invested in AI to enhance the recycling process to reuse most bottles collected. Under its "World Without Waste," the company promised to collect a bottle or a can back for every bottle or can sold by 2030.

Winnow Solutions. Winnow Solutions is a technology-based firm that uses artificial intelligence and delivers food waste management systems that will help handle the food that ends up waste in commercial kitchens. It follows the food that a kitchen wastes and what is wasted, giving insight into reasons why waste happens. In this respect, kitchens will be in a position to take actions towards minimal waste being generated. To date, several restaurants and hotels have managed to significantly minimize food waste while saving much through incorporating the technology from Winnow.

Too Good To Go. This is an application that connects consumers to restaurants and retailers selling surplus food at discounted prices. The application possesses AI algorithms analyzing data from food availability to consumer preferences, thus optimizing the redistribution of surplus food. Since its inception, Too Good To Go has successfully saved millions of meals from waste while promoting sustainability and reducing food waste.

ISSUES AND PROBLEMS HAUNTING THE FOOD INDUSTRY

The food industry is one of those that affects the world's economic system, as well as the health of the general population. Even then, problems and issues hinder their proper flow and continuity; it faces an inability to serve demands from clients. Some of various challenges faced are identified and described in as much detail below:

Supply Chain Disturbance

Complex, sometimes global supply chains support the food industry. Availability of ingredients and efficiency of transport networks are highly susceptible to several kinds of disruption: from natural disasters through geopolitical tensions and pandemics. For example, the COVID-19 pandemic has exposed several critical vulnerabilities in supply chains leading to shortages of essential products and delays in delivery. It will increase resilience through diversification in the supplier base, investments in local sourcing, and enhancement in the practice of inventory management. Building long-term relationships with suppliers and instituting risk assessment frameworks can help to reduce the impact from any such disruptions.

Food Safety and Quality Assurance

Food safety and quality need to be ensured by food industries to sustain their markets. The regulation requirements have been pretty strict; in fact, among such organizations, there are some like the Food and Drug Administration of the USA, and the European Food Safety Authority. Anyway, offending leads to punitive measures along with recall order and may even damage the reputation of a brand while major issues arise due to poor uniformity that may come directly from differences observed in the ingredients forming it and due to the use of different processes with interfering substances generally. Proper measures of quality, regular inspection, and technological monitoring by the food industry will maintain its high standards and consumer confidence through quality control.

Data Privacy and Security

More dependence of the food industry on data analytics and AI for even more personalized marketing and operational efficiency has given rise to data privacy and security concerns. Companies collect large amounts of information about consumers, including personal habits and purchasing trends, so their data needs protection against being compromised. Other key reasons why companies need to regulate the collection of data include data protection laws, such as GDPR and CCPA, that ensure that consumers trust the operations of the company. Recently, ransomware attacks targeted the food industry to disrupt operations or steal sensitive information. Risks can be minimized through investment in cybersecurity, security audits regularly, and the development of incident response plans.

Sustainability and Environmental Impact

Food industries continue to face increased pressures by consumer concerns of the more sustainable ways of executing their operations. Their business practices have raised unsustainable sourcing, reduction of waste, and minimal carbon footprints as concerns. Companies have to invest in eco-friendly packaging, waste reduction, and sustainable agriculture to fulfill consumer expectations as well as legal requirements. Climatic change has also affected food production through reduced crop yields and a decline in food security. Food companies havc to adapt to these changes by investing in resilient agriculture and alternative sources of food.

Changing Consumer Preferences

Consumer preferences are changing at a fast rate and is driven by health and wellness, convenience, and technology. Consumers are getting healthy, and such products that align with their dietary preferences and health aspirations are in increased demand-for instance, plant-based diets and organic food. This shift has brought both an opportunity and a challenge to food companies, which makes them conduct constant research about the markets and innovate products. Another change in consumption pattern, because of demand, is the need for convenience. Customers prefer meals that are already ready to be eaten, meal kits, and online grocery shopping. There is a need to apply technology in improving customer experience, efficient operation, and meeting higher convenience demands.

Labor Shortages and Workforce Management

The food industry is characterized by labor shortages. More common are the ones related to labor shortages within the food processing and logistics segments, as well as the service-related segments. Some factors that have driven the challenges above include low wages and poor working conditions. Other industries compete too. In terms of competing for the most valuable employee talent, a better working environment, wage packages that compete with those in other organizations, and training/development to sustain this will make companies able to attract and retain employees. Most have turned to automation and technology to get through the problem. Though it's more efficient, it's a capital-intensive process, leading to a displacement of workforces. The industry needs to strike a balance between automation and the workforce.

CONCLUSION

AI now becomes a boon for the food industry by changing the operational processes and bringing big advantages across several dimensions. Consumer demands are always on the move with the population of the world and increase pressure to provide efficiency, safety in food, and sustainability. Artificial intelligence can be a means of assistance toward the challenges brought about by such changes while food production, processing, and delivery are changing.

It has maximized production efficiency. AI has, for instance, optimized scheduling and resource usage for food makers to produce a right amount at the right moment to meet requirements with minimal wastes. For example, predictive analytics can use data on historical sales and market trends to predict when demand will reach a certain milestone and enable an organization to review its production timetable. This is a skill that will save the environment while cutting down the cost of overproduction, one of the primary factors in the food waste challenge currently facing society. Additionally, AI will track and trace products from the farm to the table in real-time. This will be very important for food safety, as companies are able to quickly identify and correct issues such as contamination or spoilage. Through improved traceability, AI ensures quality is maintained and guarantees consumer trust, making up the market with the prime demand for food safety.

Apart from operational efficiency, AI massively enhances consumer interaction so that it becomes more personalized according to consumer preferences and behavioral inputs. For instance, a restaurant can identify previous orders and consumer preferences about their dishes so that they can offer suggestions of matching dishes according to individual tastes. This type of personalization gives rise to a level of customer satisfaction and loyalty that promotes repeat business and then translates into bottom-line improvements.

FUTURE PROSPECTS OF AI TECHNOLOGIES IN THE FOOD INDUSTRY AND ITS IMPLICATIONS ON THE FOOD INDUSTRY

The future of AI technologies in the food industry appears very bright. As AI is advancing at such a rapid pace, the applications are going to increase, and thus, more innovative solutions are going to surface to overcome such crucial issues like food safety, sustainability, and changed consumer preferences. For instance, resource optimization will be more implemented by using AI-based smart agriculture technologies. With this, in real-time, the farmer will see the status of crop health and that of the soil. This would imply more harvests with no adverse effects on the environment, directly realizing both food security and sustainability. With a great advancement in its algorithm, AI is able to predict surplus food and distribute the same to all those who might need it; hence less food insecurity and aligning itself with a circular economy as it makes full utilization of any resource by minimizing waste. State powers of artificial intelligence will enhance things in food waste management for better food systems for consumers as well as to the environment.

In return, with the rising consumer awareness on issues about sustainability, it will enable this interest in food companies through enhancements in more friendly ecological practices toward the means of production. These include optimization on the consumption of energy while the plants in processing consume energy and reductions of carbon footprints emanating from logistic transportation. Therefore, AI presents all opportunities that call for sustainable practice from industry: a way of fulfilling the double dilemma of feeding on an increasingly ever-growing population alongside the need to save the Earth.

CALL TO ACTION FOR STAKEHOLDERS

Embracing AI to make the food industry more efficient, safe, and sustainable.

It will only be possible to maximize the benefits when producers, processors, retailers, and policymakers in the food sector align themselves with AI-related technologies and an innovative culture. This will be achieved through investments in AI-related solutions, training people to work alongside the developing AI, and strategic collaboration with developers of these technologies. It would then be possible to make food stand up to resilience, efficiency, and sustainability through this process. Industry leaders must also recommend deploying AI throughout the entire supply chain. This will begin with talking about best practices and collaborative research and development discussions with regulatory bodies to ensure applications of AI keep up to standards of safety and ethics. A shared commitment to welcoming AI into the food supply chain will therefore meet increasingly growing global needs for food along with a secure, efficient, and environmentally friendly system of food available to all people of the world

It is more than just simple technological advancement; it is something of overdue evolution that the integration of AI into food production might show significant differences between the very way food could be produced and processed and the way it delivers. This modern world is complex in so many ways; the promise of sustainability is huge for consumers, business, and the environment will flourish if AI can power it into sustainability. This means that stakeholders will have benefited from a progressively competitive and conscious market if AI is on board and not a matter of choice. Now is the time to act since this is where future food relies: innovation and adaptation towards new landscapes.

REFERENCES

Ding, H., Tian, J., Yu, W., Wilson, D. I., Young, B. R., Cui, X., Xin, X., Wang, Z., & Li, W. (2023). The Application of Artificial Intelligence and Big Data in the Food Industry. *Foods*, *12*(4511). PMID: 38137314

Kumar, A., Singh, R., & Kumar, S. (2020). Application of Artificial Intelligence in Food Quality Control: A Review. *Journal of Food Science and Technology*, *57*(10), 3771–3780.

Zhang, Y., Wang, Y., & Zhang, Y. (2021). Computer Vision Technology in Food Quality Inspection: A Review. *Food Control*, *123*, 107823.

Chapter 2
Artificial Intelligence in Food and Agriculture:
Driving Sustainability, Efficiency, and Innovation

A. Muthuram
https://orcid.org/0000-0003-1393-7604
SRM Institute of Science and Technology, India

M. Sukanya
Kathir College of Engineering, India

S. Alfiya
Vel Tech Multi Tech Dr. Rangarajan Dr. Sakunthala Engineering College, India

G. K. Monica
R.M.K. College of Engineering and Technology, India

G. Belshia Jebamalar
S.A. Engineering College, India

B. Maheswari
https://orcid.org/0000-0001-9711-7915
RMK Engineering College, India

P. Girija
https://orcid.org/0009-0000-1691-8180
Panimalar Engineering College, India

ABSTRACT

AI is gradually becoming the most powerful tool in the food and agriculture industries due to the possibility of making informed decisions on various

DOI: 10.4018/979-8-3693-9884-5.ch002

processes in the supply chain. Machine learning, computer vision, and robotics are widely used in agriculture to increase productivity, optimize crop yields, food safety, and apply automatization in the process of food processing as well as provide individualized nutrition and to make the food industry more sustainable and effective. This survey aims at discussing the main domains of application of AI techniques, major applications of AI in these domains, the main obstacles to the penetration of AI in different areas and the directions of further research. The chapter focuses on how AI will impact smart agriculture and the food industry, and how it will help in solving the global problem of food security and resource management in the face of climate change.

INTRODUCTION

Motivation for Adopting AI in Food and Agriculture

The food industries and agriculture have been the center of people's existence and advancement. According to the statistics, the population of the world is estimated to reach 9 billion by 2050, and as a result the consumption of foods is expected to increase, (Khang, 2023). Satisfying this demand through conventional means is economically unfeasible and unsustainable because one encounters problems like climate change, water and soil quality issues, high labor costs, and a broken supply chain. In this regard, AI is assuming the role of a modern revolution that can help to create new value, improve efficiency and quality of food production, processing, distribution and consumption starting from sowing and harvesting crops up to the final consumption. AI solutions are efficient tools to process, interpret and use the same, followed by decision-making processes. They are particularly useful in farming as soil moisture, weather, crop, and machinery data acquired from sensors, satellites, and machines can be used to enhance crop production. AI in the food industry improves production, quality assurance, food hygiene, stock management, and through recommendation and targeted advertising (Tomar & Kaur, 2021). The use of AI is not just for improving the efficiency or effectiveness of the operation but primarily to feed the nation, reduce food wastage and preserve the environment. The governments and industries in different countries have started identifying the opportunities offered by AI in these fields. For instance, Agriculture mainly targeted Digital Agriculture, Smart Food Systems are getting more funding and researched. Besides, the increasing popularity of AgriTech and FoodTech startups proves that the use of AI is gaining pace. From self-driving tractors to artificial intelligence in supply chain marketplace, the impetus is the same: AI provides a practical and data-based solution to address age-old issues in farming and food chain business.

Importance of Automation, Precision, and Efficiency

The reason why AI is being introduced mostly in the food and agriculture industry is mainly due to automation, precision, and efficiency. Many conventional practices involve the use of labor, qualitative estimates, and one-size-fits-all management approaches to the production system. These are time-consuming procedures that are also very expensive and result in overuse of fertilizers, pesticides and water, thus causing pollution and other losses. AI and roboticization of machinery solve the problem of labor deficiency in agriculture through the use of robots and machinery to undertake monotonous and strenuous tasks like planting, weeding, harvesting, and sorting. For example, mechanical pickers that incorporate computer vision are able to recognize ripe fruits from a distance and pick them mechanically with no the need for a human operator making it faster, efficient and cheap. Precision agriculture using the help of artificial intelligence implies the application of a certain treatment for particular zones of a field or even for individual plants. This is done through data that is collected from drones, sensors, and satellites or satellite imagery. This information is then analyzed by AI algorithms to give real time information on moisture level of the soil, nutrient, pests and crop status. This enhances the utilization of inputs, crop health and therefore increased yields.

In the food industry, an application of AI is in the quality control, where machines are used to vision check on the production line to ensure any defective products are eliminated hence reducing on wastage due to human error. Apart from that, it also assists in the management of supply chain through demand forecasting, shorter delivery time, and route optimization. Customer service through the use of chatbots and recommendation systems has been a plus to consumers and customer loyalty. To sum up, the importance of AI is in its capacity to achieve more with less, to provide more food with less resources, in less time and with better precision. This is so because the industry is under pressure to feed the increasing population as well as address climate change issues.

Scope of the Survey

This survey chapter will therefore seek to give an outline of the areas that AI is applied in the food industry as well as agriculture. It ranges from machine learning up to deep learning, computer vision, NLP, Robotics, and holds all the examples of how these technologies are being applied to solve business problems in the sector.

Some of the areas of application where it can be used includes precision farming, crop yield estimation, pest and diseases identification, smart irrigation and use of robotics in agriculture. Food processing automation, supply chain management, food safety, nutrition and information systems for food consumers. This chapter not only discusses the state of the art of research in the field, but also the practices that have been implemented by companies and organizations. It also covers subjects that concern the relations between AI and IoT, blockchain, edges which can give one a broad understanding of smart farming and food systems. The discussion is backed by literature, technology trends and developments in the current society. Additionally, the chapter discusses the biochemical and information system ethical issues with regards to using AI including data privacy concern, algorithmic bias, high costs, and the absence of standard regulatory frameworks. It also points out the directions for future research that may define the further development of the AI-based food systems.

Organization of the Chapter

The organization of this chapter is designed to help the reader to follow the logical sequence of topics from the general AI technologies to the specific technologies and their consequences. Section 2 highlights the definitions of the AI technologies that are most applicable to the agriculture and food industry namely, machine learning, deep learning, computer vision and AI-integrated IoT systems. Section 3 is on the application of AI in the farming sector where it discusses precision farming, yield prediction, pest detection, irrigation and agricultural robotics. In section four, it draws the use of Artificial Intelligence in the food industry concerning food processing, supply chain, food safety and quality, nutrition, and customers. Section 5 provides the main impediments and drawbacks that limit the applicability of AI in these sectors. It is in section 6 that the author provides future prospects of AI such as its compatibility with the new technologies and how the implementation can be made sustainable. In section 7 of the chapter, the author provides a brief synthesis of the work and a thoughtful contemplation of the role of AI in the future of food and agriculture.

OVERVIEW OF AI TECHNOLOGIES: FOOD AND AGRICULTURE

The implementation of Artificial Intelligence (AI) in the food and agriculture industry depends on various significant and a wide range of technologies. All of these technologies solve certain problems and offer particular applications improving the food system being smarter, more efficient, and sustainable. This section gives brief information on the basic sets of technologies that form the basis for innovation in these domains which include artificial intelligence, machine learning, deep learning, computer vision, Natural Language processing and Robotics, Artificial Intelligence of Things.

Machine Learning (ML)

Artificial Intelligence (AI) is a broad field of study that is defined as a system's ability to perform tasks that are intelligent in nature while Machine Learning (ML) is a subfield of AI that enables systems to learn and make decisions from the data provided without being programmed to do so, (Hiremath & Ingalagi, 2025). In agriculture and food industries, the application of ML is common in prediction, identification of outliers, categorization, and decision-making tools, (Chhikara et al., 2025). In the agricultural sector, the use of ML models can predict crop yields of a certain season depending on the weather conditions, soil health, and past farming practices. Such predictions assist farmers in making right decisions regarding the time and place to cultivate or to apply manure or water. For instance, the algorithms such as support vector machines (SVM) and random forests are used in the classification of the crop diseases and the possible treatments. In the food industry, ML is used in demand prediction where the sales data and trends in the market are used to estimate the future demand. This is good for inventory control and avoiding wastage of food. There is also the application of ML in quality control where the system is trained to identify defective products from the data from the sensors and the images. Also, unsupervised learning is used to cluster foods so that they can be classified based on similarities for use in the automated retail systems. In general, ML is critical in supporting decision-making in every process since it helps in converting large data into useful information.

Figure 1. AI technologies in food and agriculture

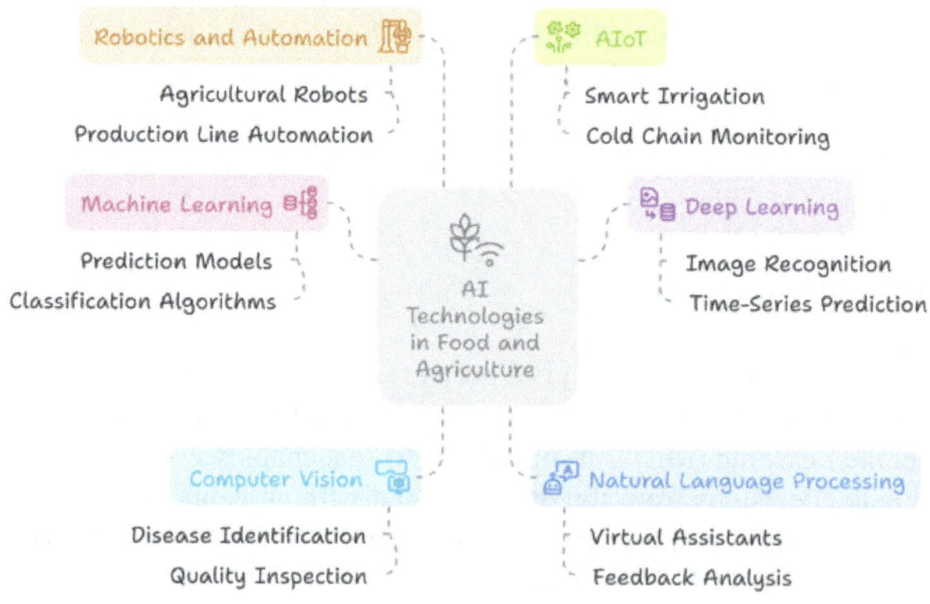

Deep Learning (DL)

DL is a subfield of AI that has a group of layers in its neural network and is rich in data representation ability, (Prabha et al., 2022). DL models are very useful in certain tasks such as image recognition, voice analysis, and sequence prediction that are important in farming and food industries (Amuthadevi et al 2025). In the field of agriculture, CNNs are utilized for image classification of crops through drones or hand-held devices for ailment identification, pest and nutrient deficiencies. These models out-compete conventional image processing methods since they are capable of learning features from the data in a hierarchical manner. Recurrent Neural Networks (RNNs), especially, Long Short-Term Memory (LSTM) networks, are widely used for time-series prediction, for example, predicting the weather or irrigation requirements. DL models are also employed in the identification of the use of land and monitoring of crops on a large-scale using satellite data. In the food industry, DL algorithms are applied in smart sorting systems used in categorizing fruits and vegetables depending on their size, shape, color and ripeness. It is also used in creating recommendation systems for nutrition apps, where a person's behavior, eating patterns, and health conditions are used to produce a meal plan. This is because the DL is capable of handling large and unstructured data such as images and videos, which are crucial in enhancing precision agriculture and automating food production line, (Siva Subramanian et al., 2021).

Computer Vision

Computer Vision is the specialty of Artificial Intelligence that allows a device to analyze and comprehend the pictures or videos captured in the real-life scenarios. It is useful in all the scenarios where image or video analysis is required, which makes it particularly important in agriculture and the food industry. In the farming industry, some applications of computer vision are in disease identification for plants, detection of weeds and determination of ripeness of crops, (Rajendran et al., 2023). These include images from mounted cameras on drones or robots of the field which are then analyzed to identify signs of disease or pest infection. These systems can cover a large area and inspect it much quicker and more efficiently than an inspector. Some robots are specially designed to pick fruits or vegetables that are ready to be harvested through the use of computer vision. This in a way enhances productivity and minimizes the use of seasonal workers in the organization. In the food processing industry, it is implemented to check quality on the production line from inspection. The cameras are used in monitoring the quality of products by checking for color, shape, size and surface defects in real time without the need for human input. Non-food product sorting machines using vision system can sort out foreign materials or even improperly packaged foods to improve food safety. Computer vision is also applied in the retail sector, for example, in shops where the customer is not required to go to the cashier's desk because the AI system identifies the products that the customer takes and charges accordingly.

Natural Language Processing (NLP)

NLP is an AI technology that focuses on the communication between computers and human languages. In agriculture and food systems, NLP helps to comprehend and analyze human communication in all its forms and build machines that can do so as well. A typical use of NLP in agriculture is the creation of virtual assistant or chatbots that assist farmers in several languages. It is able to offer information on proper farming techniques, current climate conditions, methods of controlling pests, and market prices based on natural language processing of questions. NLP is also in use in the automatic extraction of information from research papers, reports, and regulations in the field of agriculture. This assists in integrating the knowledge and helping policy makers and researchers in coming up with informed decisions. In the food industry, NLP is used in voice ordering and feedback analysis in order to engage with the customers. In this case, advanced artificial intelligence algorithms are used to study customer feedback on social networks, blogs, and messages in the customer support center to assess new trends in consumer preferences and product quality problems.

NLP is also applied in recipe generation systems where it involves the comprehension of ingredients and cooking instructions to recommend meals to the users. Therefore, NLP enables the interaction between humans and complex systems, making artificial intelligence friendly to the farming and food business without the need for technical knowledge.

Robotics and Automation

Robotic technology and automation are gradually becoming the new trend in agriculture and food processing through the use of machines that replace the human labor force. These systems are typically controlled by artificial intelligence to make decisions in this environment. Some of the uses of robots in agriculture include planting, seeding, weeding, spraying, as well as harvesting. This makes these machines to be capable of moving in fields, sensing obstacles on the field and altering their operations in accordance to feedback from the field. Robotic harvesters are expected to be able to pick fruits which are delicate in nature in a very accurate manner so that the fruits will not be damaged and at the same time, maximize the speed in picking the fruits. Drones are a kind of agricultural robotics that are employed in aerial surveys, spraying of pesticides, and farming maps. Using AI, drones can determine the state of the plants in real-time and go further, estimate the yields by evaluating the density of the plantations. In the food industry, production lines with robotic arms are used for packaging, labelling, and sorting of the products. They increase the versatility of the models and do not require reprogramming for different varieties of the product. In warehouses and logistics, robotic systems in specific the order picking and material handling have been known to work faster and with a higher precision. Combining it with robotics enhances output, minimizes the chances of errors, and develops robust systems to run continuously throughout the day.

Internet of Things (IoT) and AI (AIoT)

IoT is defined as the connection of sensors and other devices that are used to gather and share information. When this is done, the integration is called the AIoT (Artificial Intelligence of Things) which makes it possible to carry out intelligent decision making at the edge of the network, (Anusha et al., 2024). In the smart agriculture, IoT sensors include soil moisture, temperature sensors, humidity sensors, pH sensors, and plant growth sensors. It also uses this information to determine irrigation, fertilizing, and pest control timings which enhances the use of resources and production, (Sudha et al., 2023). For instance, in the smart irrigation method, data acquired from the sensors and data from the advance weather prediction would help in irrigating crops only as required with no wastage, (Sudha et al., 2024). Likewise, AIoT systems can also detect changes in the conditions of greenhouses and are able to control lighting, temperature, or ventilation. In the food industry, active IoT devices are utilized in tracking cold chain logistics so as to guarantee optimum storage and transportation conditions to foods. It also prevents food spoilage and ensures food safety since the AI algorithms can identify possible failures in real-time. In addition, AIoT systems are useful in the aspect of traceability and transparency, (Ponnuviji et al., 2025). Starting from production, through transportation to consumption, the food can be monitored through sensors and analyzed with the help of AI in order to guarantee that it meets all the required standards and to increase consumers' trust.

AI APPLICATIONS IN AGRICULTURE

The fourth industrial revolution continues to revolutionize the agricultural industry through the development of intelligent systems. Today's technological advancements in applying AI for the agriculture industry, which include; Machine learning, Computer vision, and IOT has ensured that the agricultural industry switches from the traditional agriculture to smart agriculture. This section focuses on the major areas of the application of such technologies in agriculture that includes Precision farming, Crop yield prediction, Pest and disease detection, smart irrigation and Agricultural Robotics.

Figure 2. AI applications in agriculture

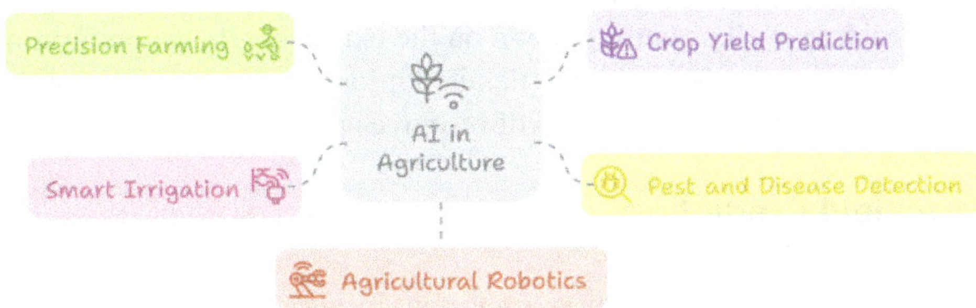

Precision Farming

Precision farming or precision agriculture is a farming management concept that uses information technology, artificial intelligence and precision agriculture technologies to control the farming activities with high precision. The best treatment has to be given at the right time and at the right place to avoid wastage of resources as much as possible while maximizing output, (Prabha et al., 2024). Soil monitoring is one of the key aspects of precision farming that is intended to improve crop production. Machines and sensors placed in the soil to monitor the nature of the soil such as pH level, moisture level, temperature and nutrient level. These data are processed in real time to ascertain the exact need of the soil through a machine learning algorithm. This assists the farmers to decide when and how much of fertilizer or water to use for their crops, (Prabha et al., 2021).

Crop health prediction is another important application of AI along with computer vision that can identify the stress in plants due to nutrient deficiency, pests, or diseases. Used in drones or robotic systems, field images are obtained with cameras and through deep learning the first signs of early deviations are identified, (Ganesan et al., 2024). This ensures that precaution is made in advance to avoid incurring heftier losses in the future. fertilizer and irrigation management are based on recorded data as well as current data collected from the plants. Some of the areas it covers include, the recommended type and amount of fertilizer required depending on the particular crop, crop cycle and nutrient composition of the soil respectively. Likewise, the smart use of irrigation systems makes sure that water is delivered to the right place at the right time, hence improving the use of water as well as cutting down on the expenses. Precision farming therefore equips the farmers with detailed information to make more informed decisions on the right use of resources and sustainable farming.

Crop Yield Prediction

Yield estimation is extremely important in agricultural management and food chain planning. Conventional approaches to forecasting are usually the use of statistical models and may not consider the changes in the environment. Automated models of forecasting are more accurate as compared to manual ones. By applying crop yield prediction models, these algorithms can help in analyzing yield records, soil type, seed type, amount of water supplied to the crops and satellite imagery to estimate the yield. These models learn from the data to find out patterns and factors that have an impact on the performance of crops. One of the major factors that should be considered when using AI to make predictions is the inclusion of weather data. Due to the fact that the climate influences the health and maturation of crops, AI models consume information regarding rainfall, temperature, humidity, and wind velocity. For example, for analyzing sequential data such as weather data, there is LSTM (Long Short-Term Memory) that can predict the yields that an upcoming drought will have. Crop yield prediction is not only useful for a farmer to plan his activities but also for the government and other agribusiness to determine its import-export policies, storage, and food security policies.

Pest and Disease Detection

This is one of the worst problems that affect agriculture since pests and plant diseases are known to wipe out an entire farm if not detected in good time. However, AI has a great potential for early detection and monitoring of these risks and thus reduce them greatly. Among the applied methods of pest identification and diseases, there are computer vision and image classification. Farmers or automated systems take pictures of the plant leaves and stems and feed them to the AI models trained on the large datasets of diseased and healthy crops. This is because these models can identify early signs of a disease that may not be distinguishable by human sight. This capability is further improved by drone surveillance as it can help in monitoring large fields. Mounted on platform that are high-resolution cameras alongside multispectral sensors and the fields are photographed. This data is processed by AI to identify changes in colors, lesions and other patterns that would suggest the presence of pests or infections. AI can also create heat maps that will help identify areas that are heavily infected so that the farmer does not have to spray the entire field with pesticides. It reduces the use of chemicals, cuts on input cost, and helps in the conservation of the environment. Also, the early diagnosis of diseases isolates and treats the sources and reduces the spread of the diseases.

Smart Irrigation and Resource Management

Water is one of the most important and at the same time a fast-dwindling resource in the agriculture industry. Intelligent systems in irrigation benefit the farmers by conserving water and increasing the productivity of food crops. Smart solutions for irrigation are built on the basis of IoT systems with the help of Artificial Intelligence. Moisture content in the soil is monitored by probes that are located in the soil and the weather conditions are also taken into consideration. AI studies this information and decides when and to what extent the irrigation should be done. This helps in water management whereby crops are watered to the right amounts so that there is no excess watering or lack of it due to drought. Water usage analytics also help in long-term planning on the usage of the resource. AI can determine how much water is used in summer, winter, and other seasons, how much water is used in different fields, and where the problem of high consumption is. These findings can help the farmers to change their irrigation regimes and replace them with more efficient approaches such as drip or sprinkler irrigation.

Also, other inputs like electricity for the pumps and fertilizer usage can be managed by AI. The integration of sensors and Machine Learning can therefore forecast nutrient runoff and minimize on the instances of over-fertilization and environmental pollution. The integration of AI in the management of resources not only improves the sustainability of agriculture but also cuts the expenses that farmers incur immensely.

Agricultural Robotics

Robotics is slowly becoming the new trend in farming as it transforms how traditional farming activities are conducted. Including AI in these robotic technologies enhance the functionality of perceiving, analyzing as well as acting independently making farming business more productive. Harvesting robots are a prime example of this since they are used to pick fruits from trees such as grapes, strawberries, oranges, and others. These include the use of machines that are fitted with computer vision systems that help in detecting ripe fruits or vegetables and in a gentle manner pick them without causing any damage. Robot harvesters do not get tired and can work for 24 hours at a time and are consistent in the quality and speed of their work, which is very important during harvest period. Weed removal and planting are some of the other activities that are well executed by agricultural robots. This has made it easier to use mechanical or chemical control in weeding because the AI algorithms differentiate between crops and weeds. It also cuts on expenses that would have been used to purchase herbicides thus promoting environmental conservation. Automation in planting requires using robotic technology whereby the field is mapped and planting is done with accuracy in terms of position, depth, and distance. Some of these robots rely in GPS and artificial intelligence to determine the field work and decision making. Agricultural robotics therefore cuts on the reliance on human labor, increases accuracy and the scale of production. With the increasing occurrence of labor shortage in many nations, AI powered robotics can be a reliable solution for continuous food production.

AI APPLICATIONS IN THE FOOD INDUSTRY

Over the recent decades, the food industry a broad and dynamic segment in global economy has undergone massive shifts through integration of Artificial Intelligence. In food processing, AI helps to improve food quality, in supply chains, to avoid losses, and in the consumer experience, AI makes it possible to provide the best possible experience. This section focuses more on five broad categories on how AI is delivering a revolutionary change in the food industry: Food processing and automation, supply chain, food safety and quality, personalized nutrition and customer interface.

Food Processing and Automation

AI has helped the food processing industry in the following ways: Automation of laborious and time-consuming processes and enhancement of the efficiency of the processes. The first of these is quality assurance in which AI systems are used to check on the quality of products such as size, shape, texture and appearance. In traditional manufacturing environment, the main method used in quality control was inspection that was costly, slow, and subjective. In the present time, with the help of sensors and imaging devices, it is possible to incorporate the AI algorithms that can take a quick look at all the products coming in the production line. For instance, AI can identify any defects on the surface of baked products, examine the consistency of the chocolate coverings, or the doneness of meat. These are real-time systems that minimize on the production of faulty products and enhance product quality. Defect detection is one of the areas where computer vision is of paramount importance. Cameras are installed on the conveyor belts that take snaps of food items and then these images are passed to AI models developed on several thousand samples. Such models can detect such flaws as discoloration, crack, bruise, or foreign object on the surface of the fruit. In case of any defective product, they can be discarded from the production line to prevent their sale to the consumers. Also, more and more robotic arms operated by Artificial Intelligence are used to cut, sort, and pack food products. These robots can handle a wide range of sizes and shapes of the product, thus being more versatile than other machines. This increases productivity while at the same time ensuring that the hygiene standards within the food processing environment are observed.

Supply Chain Optimization

The food supply chain may be an extended one and may include several steps from sourcing of the raw materials to the delivery of the final product. AI assists with these tasks by enhancing the clear visibility and recognition of likely demand and minimizing those other complexities and drawbacks of operation of the business. AI is used in several areas, with one of them being particularly important for inventory management – predictive analytics. AI models use past performances and statistical data of sales, seasonal trends, and some particular cases like holidays or changes in the climate. This leads to determination of the right stock quantities to order, avoid excessive or inadequate stocks. For instance, supermarkets can apply AI to predict sales of certain products within certain periods of the year, so they order them in advance and do not let them go bad. AI in logistics has been very useful in improving the performance of deliveries by providing the best routes to take in terms of traffic, availability time and even fuel costs. It can also help in monitoring the conditions of the vehicles and guaranteeing that temperature-sensitive products are shipped under the right conditions. When it comes to customers, the results are higher satisfaction due to shorter waiting time and better track record, while the company benefits from lower operating expenses. Demand forecasting also helps the manufacturers to plan their production schedules in the most appropriate manner. Promotion, pricing and even competitors' strategies can be evaluated to determine their impact on demand using AI tools. Such a level of foresight will go a long way in ensuring that the supply chain meets the market demands hence less wastage and high profit margins.

Food Safety and Quality Assurance

Food safety is one of the most important concerns in the food industry and AI is used in detecting contaminants, monitoring storage conditions and compliance with the set standards. AI for detecting contamination entails the use of image recognition as well as the sensor data in order to detect contamination in foods. For instance, hyperspectral imaging integrated with AI can project and differentiate microbial infestation or chemical traces that cannot be seen with the naked eye. This is to avoid exposing the people to various complications and also to guarantee food safety in the market. It is also important to note that the development of smart packaging that includes AI and IoT technologies improves spoilage prediction. Smart packaging is the packaging that contains sensors that can change its state depending on the change in the composition of the gasses, temperature, and humidity which are signs of the freshness of the food. For this reason, this data is fed into an AI algorithm to estimate the shelf life remaining for the product. This information can then help manufacturers and retailers to focus the promotion of these products and avoid wastage. Further, the AI tools assist in ensuring hygiene and sanitation of the production environment in order to meet the agreed food safety standards. It can also monitor whether or not machinery has been cleaned and used cleaning agents, and if there are any violations of the program. In this way, AI enhances food safety measures and gives the customers more confidence in the quality and safety of the food they are consuming.

Figure 3. AI applications in the food industry

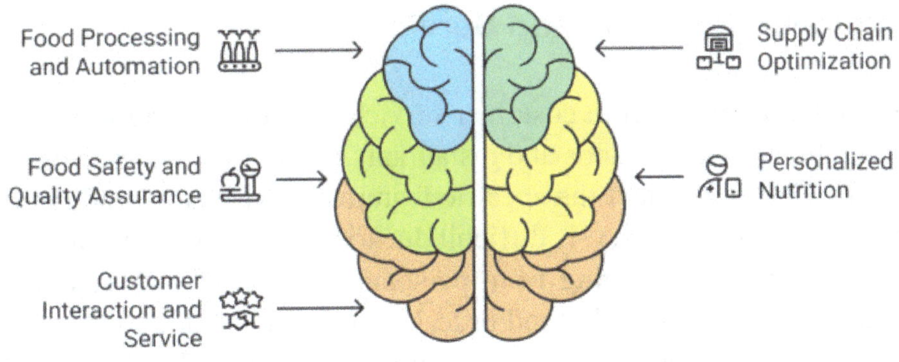

Personalized Nutrition and Diet Planning

It is also improving its presence in the field of nutrition where it offers customized diets in accordance with the health status, preferences, and other parameters of individuals. Several recommendations include age, gender, personal medical concerns, allergies, and goals and objectives to fitness when offering diet plans. The last generation of these systems relies on extensive nutrition databases and employs artificial intelligence to learn more from the users' experiences in utilizing it. For instance, a user with diabetes may get suggestions on the type of meals that will ensure that their blood sugar level is kept at the required level. This is because wearing and health applications also improve the personalization with wearable devices. Fitness and health data from wearables or others such as, smartwatches or glucose monitors are given as input to AI models that can predict caloric intake, physical activity, and metabolic outputs. It provides an opportunity for the changes in the recommendations depending on the course of the disease and the patient's condition, thus, achieving better long-term compliance and results. Apps with food delivery services and meal kits are also applying AI to provide the user with meals that are compatible with their diet, such as vegan, keto, or low-sodium diets. These systems gradually acquire information about one's preferences and offer recipes or menus that are healthy and tasty. Thus, AI acts as a mediator between research and real-life consumption, contributing to the improvement of people's diets.

Customer Interaction and Service

The integration of AI in customer service has gone a long way in enhancing the way food brands and restaurants relate with their customers. Here, the ability to select the specific features of AI, that would help in changing the food industry, is difficult due to versatile applications of AI in the food industry ranging from personal interaction with customers to smart assistance. Mobile food ordering apps for fast food restaurants are the perfect example of where chatbots can be found. These virtual assistants can entertain customer inquiries on the status of their orders or concerns on payment or refunds at any time. This way, using NLP, chatbots are capable of interpreting customer messages and providing an instant response that enables a faster response time and higher customer satisfaction. Furthermore, voice assistants in smart kitchens such as Amazon Alexa or Google Assistant are used to help the users to plan meals, search for recipes or even manage smart cooking appliances. These assistants are capable of interpreting the instructions on how to cook, measuring and make recommendations as to what could be used in place of a missing ingredient. It makes cooking easier and fun, especially to those who do not have time or have little knowledge on cooking. Restaurants and food retailers are other industries which have incorporated the use of AI in their marketing strategy. AI models also identify the buying pattern, past purchases, and preferences of customers to send coupons or suggest dishes. Loyalty programs that are linked with artificial intelligence can offer individualized rewards to the customers, which in turn improve their loyalty. Thus, AI contributes to the improvement of food business services by offering an efficient approach to customer engagement.

CHALLENGES AND LIMITATIONS

As this chapter demonstrates, the food and agriculture sectors stand to benefit from the integration of AI; however, there are several issues that arise when adopting this technology. Despite the advances in AI technology, several barriers limit widespread adoption, especially in regions with limited infrastructure and resources. This section identifies the main issues and restrictions concerning the application of AI in food and agriculture such as data quality and accessibility, high cost and accessibility for small users, ethical and privacy issues, and absence of specific domain models.

Data Quality and Availability

A major issue that arises in the implementation of AI solutions is the availability of good, pertinent, and adequate data. AI depends greatly on data for training and also the information that is used for making decisions. These are climatic data in relation to agriculture and the food chain, soil data, crop data, pest and diseases data, supply chain data, and consumer data. However, data in these sectors is unsorted, unsystematic, or stale most of the time. For instance, in many developing areas, farmers have not developed computerized records of their farming activities and records in agriculture are taken manually which is full of gaps. Weather and satellite data may also not be available or may be too coarse to support the decision-making at the village level. In the food industry, data from the processing units or logistics chains may be fragmented in various platforms or systems and may not be easily integrated and analyzed. In addition, the problem of bias and imbalance in the training data sets also have an impact on the performance of an AI model. The model may be trained using data mostly obtained from one area and crop type and this may not work well when applied to another situation. Handling and using data: Data standardization and data compatibility at the level of different systems which is highly important is almost lacking to provide the necessary basis for building adequate scalable AI solutions. This factor will remain a limitation to the full potential of deploying AI in these sectors if there are no developments in the data infrastructure, digitization practices, as well as data sharing frameworks.

High Costs and Accessibility for Small Farmers

Another challenge that has been stated is the cost of adopting the AI technologies which may be a major challenge to small scale farmers and food producers. Developing and sustaining an AI system requires a considerable amount of resources in terms of hardware, software, and networks. For instance, precision agriculture involves the use of drones, sensors, GPS, and AI among others, something that small holder might not afford easily. Thirdly, the expensive internet connection in the rural areas also hinders the implementation of the cloud-based AI applications that rely on real-time data transmission and remote monitoring, (Dhinakaran et al., 2025; Ponnuviji et al., 2025). It also increases operational costs as training of the staff or farmers in the use of the AI tools may be required since many of them may not be technically inclined or may not be literate. Also, the AI solutions that have been designed and tested in the industrialized setting may not be very relevant for the smallholder farmers' problems like low mechanization, small and sub-divided plots, or mixed crop-livestock systems. This is worse off since it also implies that the system is not customized for small-scale usage, thus lowering the value proposition for such users. Therefore, there should be a focus on developing affordable local AI solutions and offering subsidies, microcredit, or PPP to make it more affordable. External resources such as funds from the government and podia which are free and open source also have the capability to foster inclusiveness of the use of AI techniques.

Ethical and Privacy Concerns

The application of AI in food and agriculture has an implication to the various ethical and privacy concerns particularly when the technology involves data collection and retrieval. Thus, for instance, in personalized nutrition, it is possible that AI systems will utilize the data from the health apps, wearables, or genetic data to provide dietary recommendations. However, if not well managed, this data may be used for negative purposes or yield negative effects like discriminations or data breaches.

Another significant issue of digital farm management is ownership of data in agriculture. Some questions that are often asked include: who owns the data generated from the sensors placed on a farmer's field – the farmer, the equipment manufacturer or the software provider? This is due to the possibility that farmers may end up losing ownership of their own data, which in turn could be exploited or frozen in the market. AI systems also have algorithmic bias by which decision making is influenced by the set of data used in the AI system. This could be seen through crop yield predictions that are inclined to benefit some areas or populations, or supply chain choices that provide substantial preference to large buyers over small growers. Moreover, there are concerns that the use of technology in accomplishing certain tasks can result to loss of jobs such as using robotics in food processing industry. Consequently, this raises social issues that can only be managed through public policies on workforce transition and reskilling. For this reason, there is a need for proper ethical guidelines for the use of AI as well as data governance policies. There are best practices Xia et al. have outlined that needs to be adopted in these sectors comprising of transparency, accountability, and inclusiveness when it comes to development of artificial intelligence.

Lack of Domain-Specific AI Models

This is because there is a general problem with AI models in that they are not very effective if used in a specific field such as agriculture and food production. This is because the conditions that define the farming environment vary, for instance, the climate in different regions, type of soils, the crop varieties and the methods of farming. Consequently, a number of AI tools can only be applied conditionally or need to be adapted for a long time before they can be used in new environments. For instance, a crop disease detection model trained on images of tomatoes grown in green houses and under controlled conditions will not be useful when applied to tomatoes grown in the open field under the tropical climates. In the same way, AI models for food quality assurance may not be sensitive to the differences in taste, texture, or processing among the regions. This is due to the fact that annotated datasets are often scarce particularly for certain crops, pests or food products. They often need to consult with experts in Artificial Intelligence and knowledge engineers in terms of needs of the agricultural scientists and food technologists and so a lot of time and money is used in the development of such domain models. Moreover, there are other barriers such as language and culture that may influence the use of AI tools in a particular region. A chatbot created using NLP in English may not work effectively in the rural areas where there are regional languages spoken. To overcome this issue, AI solutions have to be built in close collaboration with the domain specialists, focus is to be made on local data acquisition and model training, and the frameworks themselves should allow for certain kinds of flexibility. This will also help in making sure that the solutions that are being developed will be useful to the users of the system since they are the ones who are directly involved in developing the AI. Thus, AI has a great potential for the further development of the food and agriculture industry However, there are several important issues that need to be solved. Challenges of implementing AI for smallholders include data quality, price and accessibility of technologies, ethical and privacy issues, and domain-based models. Thus, sharing these limitations and establishing synergies between the stakeholders, including the governments, academic institutions, industries, and communities, AI could be helpful in advancing and making intelligent the food systems in order to meet world's food needs, while also making the process sustainable.

FUTURE DIRECTIONS AND RESEARCH OPPORTUNITIES

Given the current advances in artificial intelligence, it is projected that the application of the technology will increase in the food and agriculture sector. Though, there is substantial applicability in precision farming, safety in agriculture, and automation, there are many more possibilities that are yet to become mainstream. This section looks at possible developments and more research that can be done to improve the role of AI in food systems — the use of AI in combination with blockchain for the traceability of food chain, the use of federated learning to achieve privacy-preserving collaboration in using AI, the use of AI that can be explained for food safety and the use of AI in climate-smart agriculture.

Integration of AI with Blockchain for Traceability

A very relevant avenue of development is the application of AI with an improvement of supply chain transparency and traceability issues with the help of Blockchain. Globalisation of food is another significant factor where food is produced in one country and is sold to another through several middlemen that include producers, processors, distributors, and retailers. This might not only complicate the tracking of the source of products, food safety or holding ethics and ecological tests.

Blockchain is a distributed database where each transaction or event in the supply chain could be recorded and validated. This information is valuable when incorporated with AI to identify the weaknesses, fraud and even make correct decisions in the supply chain. For instance, it is possible to use AI to monitor data from blockchain-integrated sensors for cold chains to determine spoilage or suggest new routes based on real-time data.

Moreover, AI algorithm based smart contracts can perform quality assurance checks or even issue warnings when there are deviations for quick and accurate response and compliance. In the situation where food products have to be recalled due to contamination, AI-blockchain can quickly identify the batches and their sources that have been contaminated, and this will reduce the time taken to respond to the issue and the risk to the public health.

In this context, possible areas of interest to explore for creating more relevant models for the future are the development of common platforms that require more compatibility and integration between AI and blockchain, as well as the scalability challenges and standard protocol, validation of data and smart contracts execution among the stakeholders, (Balakrishnan et al., 2024).

Federated Learning in Agriculture

Therefore, in the period of enhanced application of ITS in the agriculture sector, issues of data ownership become of paramount importance. Some of the concerns that make farmers and agribusinesses reluctant to share their data are the potential misuse of the data or fall into the wrong hands and or the fact that they will be exposed to competition. On the same note, the models that are developed must be fed with considerable amounts of good quality data from various sources as a way of enhancing accuracy and resilience. FL is an innovative AI paradigm that enables machine learning models to be trained on a number of devices or servers but the data remain on the participant's device, (Sudha et al., 2023). In contrast to the architecture of the data aggregation, federated learning allows each participant a farm, a cooperative, or a processing unit to train a model locally and to exchange only the updated parameters, or gradients. These are then summed up to enhance a worldwide model. In agriculture, it can maintain data ownership and data privacy while at the same time facilitate the building of models. It also helps in reducing the bandwidth and increases the compliance with the regulatory requirements of data protection. Some use-cases of FL include disease detection models where farmers upload crop images to train an accurate pest detection system without sharing their data or yield prediction models where farmers contribute data from their regions without exposing environmental data to other regions. Further studies could focus on improving the efficiency of federated learning for agricultural data, reducing the cost of communication between participants and handling issues of fairness and performance between participants with different quality and amount of data.

Explainable AI in Food Safety

As AI systems are used for quality and safety of food, contamination identification, and compliance, the demand for explainable AI (XAI) grows. Most of the conventional AI models, especially deep learning networks, are 'black-box' systems that give outputs or classifications without any rationale that a human can comprehend. Such lack of openness can be fatal for trust, particularly in highly sensitive spheres of activity such as food safety. Interpretable AI means that the results of the AI decision-making process can be explained to the involved stakeholders like the food inspectors, processors, and consumers as to why a particular outcome was arrived at. For example, if an AI system identifies contamination of a batch of packaged food, then explainable AI can point out the readings from the sensors, or the appearance of the food products, or the conditions under which the batch was processed. This is important both for legal reasons and in order to encourage responsibility and constant enhancement to the food processing industry. The models can be used for explaining the cause of safety breaches and in advising on preventive measures to be taken. Secondly, food safety entails data of different forms, such as temperature records, chemical data and images, and microbial tests, which further complicates interpretability. More research has to be done in order to integrate rule-based reasoning with learning capabilities and in order to devise methods of presenting such explanations to laymen in a comprehensible manner. Future research can also examine how XAI can enhance the use of artificial intelligence for audit trials, ensure customers trust food labels, and how to create learning models for quality assurance of food products, (Jebamalar et al., 2023).

AI in Climate-Resilient Agriculture

Climate change is one of the biggest drivers of change affecting agriculture globally through events such as droughts, floods and pest attacks and change in the growing cycle seasons. The future of AI is that it would be a passion for smart and innovative farming methods to deal with climate shocks and become a controller of climate smart agriculture. One of them is the climate-smart advice system; this may incorporate the climatic data, real-time climate prediction, satellite imagery, and soil conditions to give farmers professional recommendation on how best to engage in farming. They can be used to decide the right time to sow, which crop to plant, or when to water the crops depending on the climate forecasts. It can also be used in simulating the occurrence of some events such as heat waves or cyclones and the effects they are likely to cause on crops or disruption of supply chains so that necessary precautions can be taken. Incorporated with GIS, AI can determine climate risk at regional or at farm levels to facilitate climate insurance and disaster risk management. In addition, the AI breeding can also enhance the breeding of crops that can withstand drought, pests or diseases, or produced in marginal environments. The environmental factors, assessed through the genomic information, can inform the machine learning models to pinpoint the genetic features associated with climate tenacity, and enhance the selection. There are some suggestions which include the following: Future work on this should therefore entail developing region-specific AI models that take into account the particular context of the geographical location; Developing robust climate data infrastructure; Making climate AI technologies accessible and affordable to smallholder farmers. Therefore, it suggests that there are several areas where AI could yet have a bright future in the world of food and agriculture, address issues like traceability, privacy, explainability, and climate change. Indeed, the application rates and developments made in sustainability, efficiency and inclusiveness of food systems by affirming the AI techs will be remarkable with the help of enhancements in other such allied innovation frontiers like blockchain, federated AI learning, and genomics. But to achieve these benefits, there must be a combination of cross-sector cooperation, ethical leadership, and engagement of all the stakeholders ranging from farmers to food businesses, tech developers, and policy makers. Research and development aimed towards the future directions in these areas will thus provide a forceful and wise structure for the agri-food system in the future decades.

CONCLUSION

Artificial Intelligence (AI) has been identified as the revolutionizing tool in redesigning the food and agriculture industry. In every step of the food supply chain process, artificial intelligence is changing the conventional ways of doing things by incorporating smart ideas of automation, accuracy, and analytics. The breakthrough of applying the techniques of machine learning, deep learning, computer vision, natural language processing, and robotics has achieved a better productive enhancement of operation, means, efficiency, and effectiveness. As described in this survey, AI has been useful in agriculture and the food industry by increasing efficiency, minimizing losses, maintaining quality and satisfying consumers. In the agricultural industry, AI has brought precision farming by monitoring the soil, predicting the diseases, and irrigation. Advanced machine learning systems are highly effective when it comes to crop yield predictions by utilizing weather and environment variables, and computer vision coupled with drones applying pest and diseases identification practices. Agricultural robots are gradually replacing manual labor in farming activities including weeding and picking of crops so as to increase production. On the same note, AI integrated IoT systems are used in smart irrigation and environmental monitoring processes that are conserving resources. In the food industry, AI is bringing changes in the processing and automation of the system where quality and defects are checked on an ongoing basis. Today supply chain operations are stronger and more adaptable due to predictive analytics and artificial intelligence in planning of supply chain. Consumers are being served by AI systems that can identify contaminants and deterioration factors and even help in innovative packaging. In addition, the nutrition and customer services are being revolutionized by the AI recommendation engines, chatbots, and voice assistants that supply the health-conscious customers with personalized food solutions. There are promising developments in the field where the researchers are still working on combining with the blockchain for ensuring a transparent traceability system, federated learning for privacy-preservation and improving collaboration in food safety, and explainability to enhance the trust in AI systems. AI is also expected to have an important role in making climate-smart agriculture as well as food security in the face of the changing climate. To sum up, AI is not only complementing the conventional agricultural and food processes but is also promoting transformation in those processes that will make the food system intelligent, safer and sustainable in the global market. Nevertheless, there are a lot of opportunities to overcome difficulties connected with data, costs, and accessibility. With the proper ethical approach and public involvement, AI will be a valuable tool for addressing the future challenges of feeding the world population.

REFERENCES

Amuthadevi, C., Banu, E. A., Kumar, S. S., Karthick, S., Kistan, A., & Sudhakar, M. (2025). Aeroponics Techniques for Improved Farming Using Artificial and Deep Learning techniques. In *Utilizing Aeroponics* Techniques for Improved Farming (pp. 81-118). IGI Global Scientific Publishing.

Anusha, S., Ezhilvandan, M., Gayathri, S., Sivaraman, V., & TP, A. (2024). Emerging trends in edge AI for Industry 4.0 and 5.0: Technologies, applications, and challenges. AI Innovations in Service and Tourism Marketing, 211-228.

Balakrishnan, C., Anish, T. P., Nithya, T., Sheena, B. G., Maheswari, B., & Subramanian, R. S. (2024). Revolutionizing Trust Manufacturing Integrating Real-Time Data and AI Innovations for Enhanced Efficiency. In Industry Applications of Thrust Manufacturing: Convergence with Real-Time Data and AI (pp. 74-97). IGI Global.

Chhikara, H., Chhikara, S., & Gupta, L. (2025). Predictive Analytics in Finance: Leveraging AI and Machine Learning for Investment Strategies. In Utilizing AI and Machine Learning in Financial Analysis (pp. 325-336). IGI Global Scientific Publishing.

Dhinakaran, D., Kumar, N. J., & Ponnuviji, N. P. (2025). Safeguarding confidentiality and privacy in cloud-enabled healthcare systems with spectrasafe encryption and dynamic k-anonymity algorithm. *Expert Systems with Applications*, *279*, 127584.

Ganesan, I., Ponnuviji, N. P., Kumar, A. S., Nithya, M., Jambulingam, U., & Lalitha, S. D. (2024). Real-Time Event Detection and Predictive Analytics Using IoT and Deep Learning. In Industry Applications of Thrust Manufacturing: Convergence with Real-Time Data and AI (pp. 1-41). IGI Global Scientific Publishing.

Hiremath, V. R., & Ingalagi, S. (2025). Optimizing Digital Marketing Strategies With Machine Learning and Deep Learning Innovations. In Impact of Digital Transformation on Business Growth and Performance (pp. 175-204). IGI Global Scientific Publishing.

Jebamalar, B., Sudha, K., Maheswari, B., Nalini, M., Dinesh, M. G., & Subramanian, R. S. (2023, May). Collaborative filtering for bargaining and customer feedback in e-commerce. In 2023 7th International Conference on Intelligent Computing and Control Systems (ICICCS) (pp. 1804-1808). IEEE.

Khang, A. (Ed.). (2023). *Handbook of research on AI-equipped IoT applications in high-tech agriculture*. IGI Global.

Ponnuviji, N. P., Nirmala, G., Kokila, M. S., & Priyadharshini, S. I. (2025). Optimizing Image Retrieval in Cloud Servers with TN-AGW: A Secure and Efficient Approach. Journal of The Institution of Engineers (India): Series B, 106(2), 459-473.

Prabha, D., Maheswari, B., & Aswini, J. (2022). Wrapper-Naive Bayes approach to perform efficient customer behavior prediction. In Innovative Data Communication Technologies and Application [Singapore: Springer Nature Singapore.]. *Proceedings of ICIDCA, 2021*, 17–31.

Prabha, D., Subramanian, R. S., Dinesh, M. G., & Girija, P. (2024). Sustainable Farming Through AI-Enabled Precision Agriculture. In Artificial Intelligence for Precision Agriculture (pp. 159-182). Auerbach Publications.

Rajendran, T., Ramana, T. V., Nalini, M., Subramanian, R. S., Devi, D. C., & Anuradha, M. (2023, July). Paddy Plant Leaf Disease Classification based on Artificial Intelligence. In 2023 4th International Conference on Electronics and Sustainable Communication Systems (ICESC) (pp. 1002-1009). IEEE.

R.S. Prabha, D., Aswini, J., Maheswari, B., & Anita, M. (2021, February). Alleviating NB conditional independence using multi-stage variable selection (MSVS): banking customer dataset application. In Journal of Physics: Conference Series (Vol. 1767, No. 1, p. 012002). IOP Publishing.

R.S. Sudha, K., Ramana, K. V., SivaKumar, S., Nithyanandhan, R., & Nalini, M. (2023, February). Hybrid Variable Selection Approach to Analyse High Dimensional Dataset. In 2023 7th International Conference on Computing Methodologies and Communication (ICCMC) (pp. 1489-1495). IEEE.

Siva Subramanian, R., & Prabha, D. (2021). Optimizing Naive Bayes probability estimation in customer analysis using hybrid variable selection. In Computer Networks and Inventive Communication Technologies: Proceedings of Third ICCNCT 2020 (pp. 595-612). Springer Singapore.

Sudha, K., Ambhika, C., Maheswari, B., Girija, P., & Nalini, M. (2023). AI and IoT applications in medical domain enhancing healthcare through technology integration. In *AI and IoT-Based Technologies for Precision Medicine* (pp. 280–294). IGI Global.

Sudha, K., Balakrishnan, C., Anish, T. P., Nithya, T., Yamini, B., Subramanian, R. S., & Nalini, M. (2024). Data Insight Unveiled: Navigating Critical Approaches and Challenges in Diverse Domains Through Advanced Data Analysis. Critical Approaches to Data Engineering Systems and Analysis, 90-114.

Tomar, P., & Kaur, G. (Eds.). (2021). *Artificial intelligence and IoT-based technologies for sustainable farming and smart agriculture*. IGI global.

Chapter 3

Machine Learning for a Smarter Food Industry: Trends, Challenges, and Innovations

M. Anita

SRM Institute of Science and Technology, India

N. Subhash Chandra

CVR College of Engineering, India

N. Pavithra

Jeppiaar Institute of Technology, India

K. Bhavani

Panimalar Engineering College, India

P. Aparna

https://orcid.org/0009-0007-7510-2561

RMK College of Engineering and Technology, India

T. Thilagam

https://orcid.org/0000-0002-7657-0726

Vel Tech Rangarajan Dr. Sagunthala R&D Institute of Science and Technology, India

V. Nivas kumar

https://orcid.org/0009-0003-6080-3999

Jawahar Engineering College, India

ABSTRACT

The food industry is undergoing a significant transformation driven by the integration of Machine Learning (ML) technologies. This survey chapter explores the diverse applications of ML across key areas such as agriculture,

DOI: 10.4018/979-8-3693-9884-5.ch003

food safety, quality control, supply chain optimization, and personalized nutrition. We categorize widely used ML models, analyze domain-specific challenges, and highlight recent advancements including federated learning, explainable AI, and multimodal approaches. Furthermore, we discuss critical issues such as data privacy, model transparency, and scalability. The chapter also identifies emerging trends and proposes future research directions aimed at building sustainable, intelligent food systems.

INTRODUCTION

Motivation

The food industry is among the significant and multifaceted industries across the world and has the responsibility of serving the citizens, providing quality food, and contributing to the economy, (Bhatt, 2024). The world population is rising day by day and the level of consumers is rising with higher expectations of quality, safety, and efficiency which put pressure on the food industry to work harder and better. Old-fashioned techniques of handling food systems such as employment of manpower, linear production and decision making, and fixed structures are less effective in the modern world. This has given a reason to focus on the technological advancement that can provide smart, optimal and flexible solutions. Consequently, Machine Learning (ML), a subfield of Artificial Intelligence (AI), has risen to the acknowledgment as an innovative tool that has the potential of revolutionizing the food industry when it comes to production, processing, distribution, and consumption. Machine learning algorithms can identify the relationships between structured and unstructured data and thus, can predict, improve processes, monitor for abnormalities, and even make decisions on their own. For instance, it may be used in recognizing diseases in crops from images, in predicting the demand for products, in detecting food contamination and in managing the food wastage through proper inventory control. The inherent nature of variability, constant dependence on environmental conditions, perishable food items, and complicated distribution networks in food systems make the application of ML all the more suitable. In addition, the growing amount of data from sensors, cameras, satellites, social media, and IoT devices can be used to train and deploy the ML models to solve different real-world challenges. The increase in the technological integration in this industry has created opportunities of advancing innovation in each and every link of the value chain. Therefore, the incorporation of ML in the food industry is not just an innovation in technology but a necessity for a business to thrive. It claims to enhance productivity, quality, and profitability as well as to make the industry more sensitive to changes in the market conditions, legislation, and environmental standards. Hence, there is a rising demand for research that maps the current and future state of ML applications in the food sector, which is the rationale for this survey chapter.

Objectives of the Survey

The purpose of this study is to conduct an empirical analysis of the current state of survey on applying ML models in the food industry. Since the use of ML techniques is rapidly expanding and these techniques can be applied in various domains, it is crucial to present the state of the art in a clear and systematic manner to help researchers, practitioners, and other stakeholders to define further development opportunities. The first purpose of the survey is to classify the various ML applications in accordance to the segments of the food industry. It applies to quality assurance, food safety, supply chain, crop and livestock, consumer analysis, production, and sustainability. Therefore, when classifying the applications, the survey intends to show the extent of ML and how various subdomains use various forms of data and models. Secondly, the chapter aims at identifying the main ML models and approaches that are used in the industry. Some of them are the traditional algorithms such as the Support Vector Machines (SVM), Decision Trees, and k-Nearest Neighbors (k-NN), some of the modern algorithms are Convolutional Neural Networks (CNNs), Recurrent Neural Networks (RNNs), Reinforcement Learning agents, and some of the combined algorithms. The survey focuses on the types of these models, the ways they are chosen, trained, and tested in different food-related tasks, and the benefits and drawbacks of using them, as well as measures of their efficiency. Strengthening further, the outcome of the research activity is another crucial goal, which includes the general trends, challenges, and future prospects, which remain unsolved. Despite the enormous potential, there are certain challenges regarding the application of ML in the food industry. Some of the challenges that are still present include lack of data, interpretability of the model, integration of the system with the existing systems, compliance with the regulatory frameworks, and scalability. This chapter has attempted to discuss some of these barriers and the potential given by novel approaches like Explainable AI (XAI), Federated Learning, and combination with the IoT and Blockchain. At last, the survey aims at giving the strategic implication and the vision of the future. This encompasses the potential areas of further research in the academic field, possible ways of implementing ML in industries engaged in food systems, as well as the policy measures that can contribute to the creation of a proper and beneficial environment for the application of ML in food systems. As such, the chapter provides a fair and comprehensive view of how one can use ML to build a better, stronger, and more sustainable food system.

Organization of the Chapter

To enhance readability and provide a coherent flow, this survey chapter is organized into the following logical structure. The second section of the chapter is "Background and Fundamentals," in which the authors introduce the fundamentals of ML and the type of data that is most commonly used in the food industry. This section provides the necessary background to the different technologies and issues described in this chapter to audiences from different backgrounds. Section 3, "ML Applications in the Food Industry," is the most important part of survey and provides brief information about how ML can be applied in different domains. These include quality control, food safety, agriculture, consumer analysis, manufacturing and sustainability. For each subsection, there are use cases, models, types of data, and research findings or industrial case studies presented. Section 4 concludes the chapter and discusses a number of challenges and limitations that affect the application of ML in the food industry. Issues discussed are the ethical use of data, the possibility of understanding complex models, the necessity to make decisions in real time, and the regulations of the food safety agencies.

BACKGROUND AND FUNDAMENTALS

Overview of Machine Learning

Machine learning is a branch of artificial intelligence that allows systems to develop solutions to new problems based on previous experiences and information fed to the systems. Machine Learning models analyze data to find relationships, forecast and improve on an aspect in question. The three categories of ML are supervised learning, unsupervised learning, and reinforcement learning that can be applied to different tasks and data (Boopathi, 2024).

Supervised learning entails the training of a model on a training set in which each instance is provided with an output or target variable. The purpose of the model is to find the relationship between the inputs and the outputs and be able to predict the output of any input data not used in the training process. This type of learning is widely used in the field of food industry for the classification of defective and non-defective products, yield prediction and modeling of customer behavior. In the category of the supervised learning techniques, Support Vector Machines (SVMs), Decision Trees, Random Forest, Neural Networks are some methods.

As for the second type, unsupervised learning is the learning process that is applied to working with data that does not have any labels. The model attempts to discover a structure or hierarchy within the data that it is not immediately apparent. This technique is useful in grouping similar food products, categorizing customers, or identifying the anomalies in quality assurance applications. Popular algorithms in this case are K-Means Clustering, Hierarchical Clustering, and Principal Component Analysis (PCA). Reinforcement learning (RL) is a type of learning where an entity called an agent aims at achieving goals through its sequence of actions and decisions in an environment. It is rewarded or penalized in some form or the other, which enables it to make appropriate modifications to its strategy in the future. In the food manufacturing industry, RL can be used for setting up the right parameters for production, planning out the tasks that need to be done or even organizing the resources to be used in food manufacturing. Although RL is more complicated and time-consuming, it is more useful in the case of dynamic and multi-stage decision-making problems. Besides these categories, deep learning is a modern trend of ML that utilizes deep neural networks especially for the problems that need intricate pattern recognition like image classification, speech recognition, and natural language processing, (Pajila et al., 2023). CNNs are used in image data for instance to flagging food quality or identifying pests while RNNs and its derivatives like LSTM are used in time series such as in forecasting and supply chain analysis. There is a wide range of algorithms available for the implementation of ML and the selection of the best algorithm depends on several factors like the nature of the problem, the type of data available, computational power available and the interpretability of the algorithm. This decision in the food industry where accuracy of information and the credibility of the decision-making process are vital must be made prudently.

Figure 1. Machine learning in the food industry

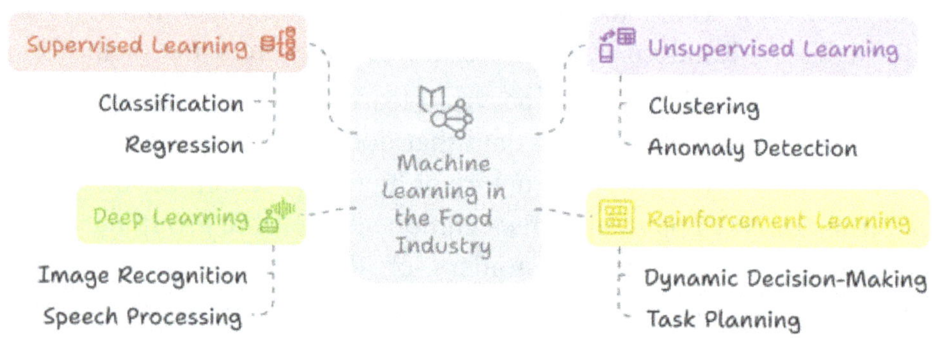

Characteristics of Food Industry Data

The data in the food industry is diverse, and it is generated in various forms from various sources. This is versatile but also poses problems in preprocessing and combining data, and in their analysis for machine learning, (Nithya et al., 2023). image data is applied in such areas as food inspection, identification of ripe fruits, crop yield assessment, and quality classification. From camera, drone or any mobile device, high resolution images pass through the ML models particularly CNNs to identify the features that define quality or otherwise. Information from sensors connected to the Internet of Things (IoT) is applied in smart agriculture, logistics and supply chain, and food chain. These include temperature, humidity, soil condition, performance of the machinery and other factors. The sensor streams are mostly in the form of time series and thus they need some sort of pre-processing to handle issues of noise, missing data and outliers. It is used by ML models in the monitoring of the systems in real time and also in the prediction of when the system will require maintenance. As much as textual data is useful in consumer feedback, safety reports, and regulatory documents, it is also a crucial factor in the analysis of such information, (Prasad, et al., 2024). Using NLP methods, it is possible to analyze such content as social networks, customer reviews, and incident reports. Opinion mining can be used to find out the public opinion, to identify emerging food safety topics and to assess the reputation of a certain product. The time-series data has its importance in the forecasting domain, including demand, price, and inventory forecasting. LSTMs, RNNs or even the Prophet model is used to capture trends and seasonality of the historical data. However, it is evident that the food industry has not fully managed to capture the value of such information. Nevertheless, there are some challenges that are frequently mentioned in the literature: Data heterogeneity: data collected from various systems or sources are generally structured and formatted with different quality. For instance, data from various sensors can come in different sampling rate or units that makes it difficult to combine them. One of the problems is that the data is sparse and contains missing values. In many small and medium-scale food businesses, it is evident that full digitization has not been realized, and therefore, little historical data can be obtained. Also, when it comes to supply chain or quality logs, manual entry may lead to some errors or some entries may be forgotten. Real-time data processing is also vital in food environments because delay in decision making may result to food wastage, compromise on food safety or even financial losses. ML systems need to not only be accurate but also fast, which means that either the models need to be small or the computations need to be done at the edge. Last but not least, Data protection and ownership are emerging issues especially in applications involving consumers. It is crucial to note that the management of customer information especially in the personalized recommendation of foods has a concern of ethics that can only be solved through secure and compliant machine learning, (Sreelakshmi et al., 2024).

Relevance of ML to the Food Industry

The use of ML in the food industry has also enormous consequences since it provides answers to numerous issues faced by the sector and creates opportunities for new practices, (Rashed et al., 2025). Food industry has its highly effective concerns as quality, efficiency, and regulatory standards. ML offers the means of fulfilling these needs with more accuracy and flexibility with the use of computational tools. In agriculture, it is used in precision farming that helps farmers to use available resources and increase production while at the same time reducing the effects on the environment. Satellite imagery as well as drone data are beneficial for disease identification, crop health conditions, as well as yield prediction with the help of the models applied. In production, it helps in process control using data mined from previous production data that would guide it in recommending the right settings for the machines, identify when a machine is likely to fail, and when the energy usage is high, (Saravanan et al 2023). This leads to increased efficiency, less wastage and therefore, cost cutting. Food safety and traceability are two more aspects where ML can be used to provide substantial value. With the help of sensor data, and logs from inspections, as well as the conditions of transport, ML can identify contamination threats, determine the cause of safety issues, and guarantee compliance with norms. Consumer behavior analysis is another important area in the current and future developments of the markets. It makes it easier for businesses to understand customer preferences, give recommendations and forecast the demand which increases customer satisfaction and reduces wastage. It is more appropriate to apply to food retailing companies, food delivery services, and e-commerce businesses. Also, ML helps in attaining sustainability goals by predicting food wastage, improving supply chain management, and aiding in carbon footprint assessment. These measures are in accordance with the international policies of the reduction of food wastage and preservation of the environment. In conclusion, it can be seen that ML is not an auxiliary tool to enhance food industry processes but an innovation that helps to create better, more efficient, and adaptive food chains. It is now more significant than ever as the industry entered into the innovation age, data generation, and automation.

ML APPLICATIONS IN THE FOOD INDUSTRY

Applying the concept of the ML has gained popularity across the food industry or nearly every subsector. With the help of ML, the entire process of production, processing, distribution, and consumption of food is changing drastically. This section discusses the different domains of food industry and how ML is applied in each of them, as well as the models and data used in each application. These examples show how the existence of ML is beneficial to optimize, address safety concerns, offer better individualization, and be environmentally friendly.

Quality Control and Inspection

Quality control is one of the first and the most widespread application areas of ML in the food industry. The conventional quality assurance techniques entail assessment of quality by human beings, which is time-consuming, costly, and imprecise. Specifically, as a branch of AI, ML is a more scalable and less subjective method than using human inspectors in combination with computer vision and sensor technologies for checking the food items during production and packaging. The most important application area of ML is image-based inspection. Images of food products are taken using cameras and the resultant images are processed with the help of ML models especially the CNNs. This is because CNNs are useful in detecting even the faint features such as bruises on fruits, discoloration in meat, irregular shapes in baked products and foreign objects. It is possible to train these models to sort products as acceptable and non-acceptable or by quality grades. For instance, CNNs have been used to sort bananas based on their ripeness or to detect mold on bread and this increases the rate of sorting lines.

Besides, there are other ways of checking and analyzing the data collected by various sensors mounted on the car. Temperature, humidity, pH, and emissions of gases in production facilities can also be checked by sensors placed within these areas. This multi-dimensional data can be analyzed using different ML models including SVMs and Random Forests. These can identify anomalies in real-time, forecast possible defects and adjust the process parameters to support the quality of the product. Using ML, quality assurance is more proactive, not reactive, and this means that money is saved, and only quality products and services get to the consumer.

Food Safety and Traceability

The safety of food is a very important issue in any country because consumers are at the receiving end and so are the regulatory bodies. This contamination may cause diseases which are a threat to the public health, recalling the affected products, and negative impact on the company reputation. ML can be used to identify contaminants and the level of risk in food products and enhance supply chain transparency. In contaminant detection, the ML models can work on chemical composition, microbial pattern or physical change. The data resulting from spectroscopic and chromatographic analysis are inputted in the supervised varieties of AI such as k-Nearest Neighbors (k-NN) or ensemble models for identification of hazardous components including pathogens, allergens or pesticides, (Gracious et al., 2023). Some applications of deep learning include identification of toxic compounds in edible oils or categorizing safe from unsafe fruits or vegetables with hyperspectral imaging, (Subramanian et al., 2024). The final area is risk prediction, which can be also considered to be rather important. Through the analysis of data on history, environment, and inspections, it is possible to predict potential safety risks using ML models. In this way, the models enable the food companies to take precautions in advance in order to avoid such occurrences. It is also used in tracking supply chain management and is highly effective when combined with IoT and blockchain. Based on the sensor data from transport vehicles, such as a temperature log, ML determines potential bottlenecks and guarantees proper storage and transportation of food items. The parameters of the environment can be used in the prediction of likely spoilage and this will enable companies to be informed in advance. Moreover, the NLP approach is also used to process safety reports, regulatory texts, and social media posts. NLP can identify the specific information needed, identify new safety issues, and provide traceability between batches and incidents. The use of NLP in ML helps the food safety systems become smarter by converting written text into useful information.

Figure 2. Ml applications in the food industry

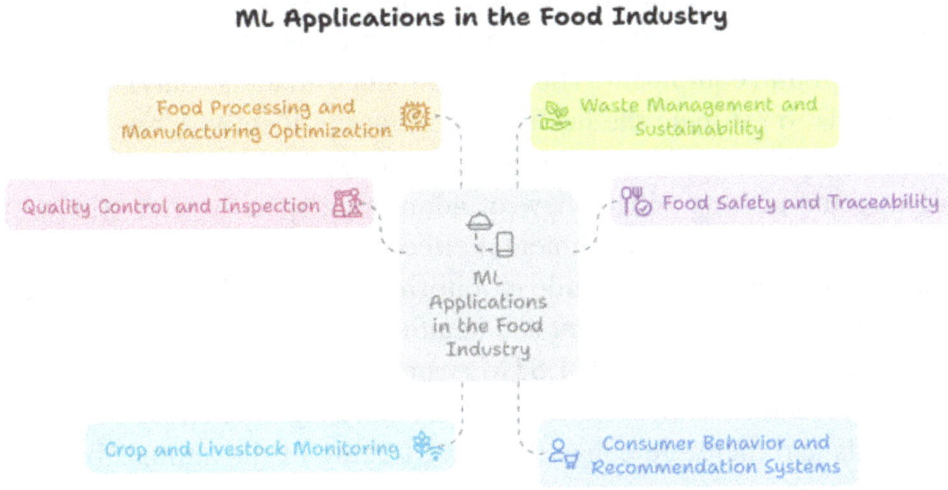

Crop and Livestock Monitoring

The earliest stage of the food value chain, agriculture has gained a lot from the new technology of precision farming as well as the incorporation of ML. Crops and livestock yield prediction, disease diagnosis, and real-time surveillance of various crops and livestock are some of the areas where ML helps the farmers to produce more with lesser resources, (Oluwaseun et al., 2024). Yield prediction is important in production, storage and distribution. There is a possibility of using ML for predicting the crop output depending on data about the soil, climate conditions, satellite images, and the previous year yield data. For this purpose, popular algorithms like Linear regression, Random forests, Gradient boosting machines are employed. Other advanced models such as deep neural networks have been employed in the prediction of the crop behavior under different climate conditions. In the case of diseases affecting plants and animals, there has been a rise in the use of ML solutions especially in the use of computer vision. In this case, farmers or drones take pictures of the leaves or the livestock to be analyzed by CNNs for signs of diseases or deficiencies in nutrients or pests. For instance, image models that learn from pictures of tomato leaves may be used to detect the first symptoms of blight and stop the disease from spreading before it affects many plants. The large majority of information that supports the concept of precision agriculture is collected with assistance of drones and satellites alongside essential ground equipment. The large amount of data is analyzed by the ML models to estimate the health of vegetation, identify stress factors, and determine the need for irrigation or fertilization. These systems enable targeted actions with a view of addressing specific sites which in the process reduce on environmental effects but increase on the returns. Other applications of ML include monitoring of livestock where it is used in tracking behavior, movement and health of the animals. Incorporation of time-series models such as LSTMs can help in monitoring the sensor data from wearable devices that indicates early signs of stress or illness in animals and therefore enhance the well-being of the animals besides reducing treatment or replacement costs.

Consumer Behavior and Recommendation Systems

In the current digital age, it is important to know and address the needs of the consumers in a business. This chapter discusses the use of ML in the prediction of the consumer behavior, recommendation systems, and demand prediction for the retail, food delivery, and e-commerce industries. Recommendation systems are one of the most visible ways in which ML has impacted the food tech industry. These systems based on recommender systems are using Collaborative filtering, matrix factorization, and deep learning for recommending recipes, products, or meals that suits the customers' taste, dietary preferences, or prior purchases, (Yamini et al., 2023).

In food delivery applications, there is the use of ML algorithms to suggest meals based on previous orders, the user's location, and the time of the day. Likewise, online grocery stores recommend related items, which in result increases sales as well as customer satisfaction. Another function where ML is quite helpful is demand forecasting. Using such factors as sales history, weather patterns, events calendar, and holidays, it is possible to build better ML models for demand forecasting. To manage inventory, staffing, and pricing flexibly, it is possible to use such methods as time series modeling (ARIMA, LSTM), regression analysis, and clustering. It is also used to assess the consumer satisfaction through their posts, comments, and feedback on the social media platforms. NLP models help in identifying the key themes, the levels of satisfaction and the emerging trends that may prove useful for business organizations to know and understand the perception and behavior of their clients. They apply to the marketing, product development and customer service strategies. By tailoring the interactions, it is possible to turn the consumer interaction from a mere exchange into an intelligent one with the help of ML.

Food Processing and Manufacturing Optimization

The business of manufacturing food products entails the integration of many sequential operations that have to be managed thoroughly due to the issues of quality and safety. It has become common to use ML for predictive maintenance, process parameters control, and resource allocation. Predictive maintenance is an application of ML that aims at predicting the likelihood of equipment failure before it happens. Temperature, vibration, current, and noise level data collected from the sensors are classified and regressed. These systems assist in effective scheduling of maintenance activities so as to avoid frequent break downs of equipment and also increase equipment's useful life. SVMs and decision trees are some of the best machine learning models that can be used for early fault signatures. Another area that has been greatly affected by the use of ML is process parameter optimization. Some of the processes that occur include mixing, heating, cooling, and fermentation, and each of these processes has variables that affect the final product quality. These settings can be learned from the historical production data to make recommendations to the ML model. Reinforcement learning is especially applied to adjust the process parameters in real-time and learn the best strategies through interacting with the system. It is also used for energy optimization, waste reduction or yield improvement with help of advanced ML tools, (Anusha et al., 2025). For example, in the case of regression models, it is possible to determine the yield of a product depending on the characteristics of the raw materials and the conditions under which the process occurs, allowing operators to take the right decisions to minimize losses. In general, ML is beneficial in creating leaner and more agile manufacturing systems as it allows to manage variability in inputs and output demands.

Waste Management and Sustainability

The concept of sustainability has become an important issue in the current food industry and machine learning is proving to be a valuable asset in addressing waste management, resource utilization and carbon footprint. Food waste prediction is one of the applications of ML that help in predicting the amount of excess or spoilage in supply chains, restaurants, and retail stores. These include time series or classification techniques that assist in determining overstocking possibilities and minimizing perishable inventory waste. For instance, a grocery chain may utilize ML in order to estimate which stocks are likely to go unsold and make changes to orders. Inventory optimization is another way through which ML boosts sustainability. With the help of sales data, shelf-life, and delivery data, it is possible to implement the ML model to manage the just-in-time inventory system to avoid overstocking and understocking. Clustering and decision-tree models help in categorization of inventory into high and low turnover items to ease on planning. It is also used to measure the amount of carbon emissions of the food products and the operations being made. Manufacturers look at different SCM information, energy consumption rates and patterns, means of transportation and sources of raw material to forecast emissions. These cues are useful for the sustainable reporting process and the process of moving towards a more sustainable approach. Also, ML is being applied in the circular food system in which food waste is converted to energy or a resource. Such models help to determine which reuse opportunities are the most valuable in helping organizations move toward zero waste.

In summary, ML provides the food industry with the required intelligence to make sustainable decisions that would be sustainable to the environment without compromising the profitability aspect.

CHALLENGES AND LIMITATIONS

Although the use of Machine Learning (ML) technologies has affected different areas of the food industry positively, the implementation is not without challenges. This chapter identifies the following challenges that affect the use of ML in this domain: Data issues and Model limitations, and Regulatory and Operational issues. It is important for both the practitioners and the researchers who are working towards the improvement of the use of ML in food systems to understand these limitations.

Data Privacy and Ethics

One of the key issues when it comes to using ML in consumer-facing food applications is data privacy and ethical issues. Most of the current ML models use the personal data such as diet, purchases, and biometrics, and health records. These datasets are valuable for nutrition plans, recommendation systems or the demand forecasting models. However, this dependency is a cause for concern since most users are unsure how their data is being collected, processed, or utilized. The food industry needs to follow the legal requirements. Failure to do so is punishable by severe penalties and may harm a company's reputation. Also, there is an increasing concern among the public for ethical AI systems that would not intrude on an individual's privacy or make use of the person's personal information without their consent. In addition to privacy, other ethical concerns are bias and fairness. This is because the data used in the development of ML systems can also contain cultural, gender, or even socioeconomic bias. For instance, recommendation systems may bring into focus some food categories or brands that are preferred by the majority in the training set while neglecting the existence of other diets. Ethical AI development therefore entails fair tests for bias, clear accountability for data utilization, and means of empowering the end user to decide how the data will be used.

Model Interpretability and Transparency

However, in many cases, especially in the food industry, safety, quality, or regulatory compliance applications, it is not only the decision made by the ML model but the reason behind the decision that matters. However, most of the high-performance ML models, especially the CNNs or LSTMs, work as a black box. This lack of interpretability presents a problem, particularly in situations that require high decision-making sensitivity such as food quality inspection, supply chain monitoring or evaluation or automatic sensory evaluation of food products.

For example, if an ML system came up with a decision that a batch of meat contains contaminants, the food safety officers need to be able to understand how the decision was arrived at in order to validate the decision and act on it. When the reasoning is not clear, it becomes hard to explain certain actions, thus the probability of false positives or missed dangers arises. Some of the solutions that have been employed in the effort to overcome this challenge are Explainable AI (XAI) methods such as SHAP, LIME and saliency maps. They can be used to identify which of the features contributed to the model's result. However, the field of XAI is still young and the main issue is to find a balance between the model's performance and its explainability. In addition, integrated models are important for trust development with the producers of the food products, the regulatory bodies, and the consumers as well. Even with the right data and algorithms, an ML system can be resisted or not meet regulations should there be a lack of openness.

Scalability and Real-Time Processing

The last major drawback of ML in the food industry is that it is not scalable and requires real-time processing. All this is due to the fact that many ML models are initially trained and tested in controlled small-scale conditions or in pilot projects. But it is not easy to expand these models to industrial levels where it involves multiple production lines, farms, or stores. Some of the use cases include dynamic pricing, spoilage prediction during transportation, and sorting of produce using robots; these need to be accomplished at low latency. This requires not only the proper algorithms for ML but also the strong architecture for the constant data input, transformation, and output. As it has been mentioned, the majority of food industry environments, particularly in the developing countries, may not have the computational facilities, fast internet connection, or data storage and management systems needed for real time analysis. Furthermore, most of the contemporary ML models are computationally intensive and may need GPUs or cloud computing services to run smoothly. The implementation of such systems can be expensive, and this may be compounded by the fact that many of these firms are small to medium enterprises (SMEs) in the food industry. These costs and resources restrict the adoption of advanced ML systems and create a digital divide among the high tech and low-tech companies in the industry. Edge computing and model compression are considered as the possible solutions to these issues, which enables the deployment of ML models on low-power devices or limited infrastructure. However, these approaches also have limitations in terms of the complexity and the accuracy of the model which has to be considered based on the application.

Figure 3. Challenges and limitations

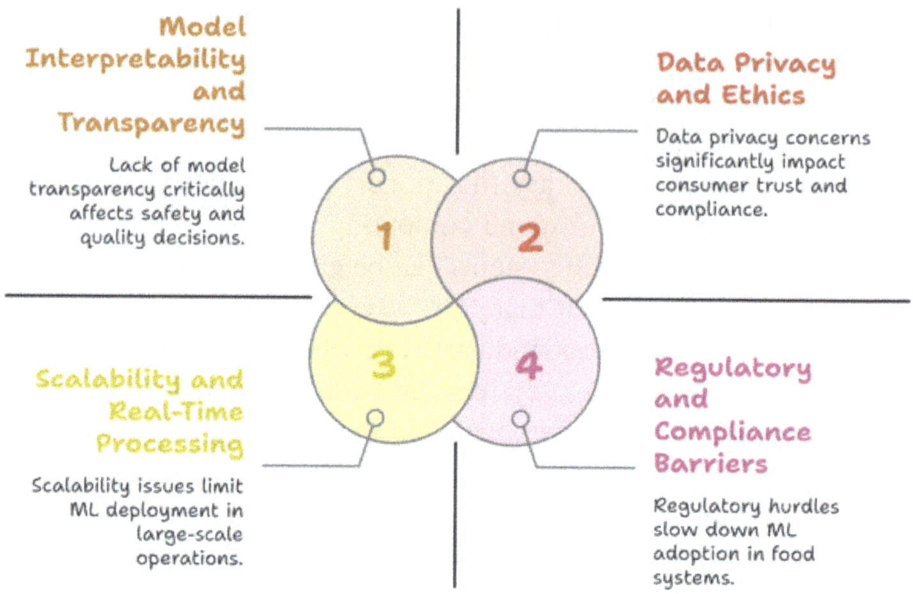

Regulatory and Compliance Barriers

It must be acknowledged that the recalls are particularly sensitive in the food industry because the sector in a way affects the population's health. Before a new technology such as ML is used in the food sector to determine food quality labelling, safety or any aspect of food traceability, the technology must meet several legal and regulatory requirements. However, current food regulations were not created with ML or AI in their framework, and this brings a lot of vagueness and uncertainty in the food regulations. For instance, if an ML-based system is able to reject a food product during the automated inspection, the regulatory bodies may require an explanation or validation mechanism which is hard to come by especially in the black box models. Likewise, the automated systems for detection of allergens, nutritional value, or origins may take time to get certified because there are no standard rules for using ML tools. This is because there are no standard international standards, hence making it even more challenging for the global supply chain of food. A model that complies with the rules of Europe may not be suitable for the United States or Asia, which narrows the company's operations. However, the adoption of ML solutions is still relatively slow because companies are still concern with the compliance or litigation issues. This implies that any innovative activity may be restrained due to lack of legal requirements and regulations. There is a significant lack of regulation that would involve risk assessments of ML systems, increase transparency and define proper ways of deployment of such systems. Efforts from the technology developers, food industry players and regulatory authorities are required to fill this gap and foster innovation with the consumer trust and safety.

Generalization Across Diverse Conditions

Another assumption that is made in the development of any ML model is that its performance will be good on data that is similar to the training data. Nevertheless, the conditions are not as stable in the food industry because they are influenced by geographic, cultural, environmental, and economic factors. An ML model trained with data from a greenhouse in Netherlands does not hold well for open field farming in India or even vertical farming in urban cities. Likewise, the options of food and quality also vary from one culture to another; this distorts the consumer behavior model or recommendation systems. This is particularly a weakness in that it may result in poor model performance, biased results or even catastrophic failures in real-world problems. For successful deployment of ML systems, it is important that the training data contains variability that is present in real life conditions. However, there is always a problem of obtaining various, good-quality data because of organizational, financial, or legal restrictions. Both, transfer learning and domain adaptation methods are proposed to overcome the limitations of the previous approach and to improve generalization. These methods enable a model to transfer the knowledge from one domain to another, with a little data from the new domain. However, the ability to maintain high performance across different conditions is still a challenge that prevents the broad use of ML in food systems. Moreover, there are still problems with learning that occurs in a continuous manner which enables models to update their parameters over time. Such systems are valuable especially where new foods, diseases, or trends that are likely to affect the consumers appear frequently.

EMERGING TRENDS AND FUTURE DIRECTIONS

Machine Learning has vast prospects as more and more technologies are being developed to revolutionize the food industry. The symbiosis of ML with such trends as Edge AI, xAI, IoT, and Blockchain is opening new opportunities to create more effective, safe, and intelligent food chains. This section discusses such trends and suggests promising further research avenues that are likely to revolutionize the food industry.

Federated Learning and Edge AI in Food Systems

Among the most recent trends in privacy-preserving and decentralized ML, there is Federated Learning (FL). In standard ML, the data collected through various sources is centralized in a server for the purpose of model training. This approach, however, is invasive of data privacy and is not practicable in scenarios where data is dispersed in many places. 'Federated learning' has been proposed to combat this problem by letting individual devices or nodes such as farms, food processing units, or retail points train the local models and feedback only the updates to a central server. This method is particularly useful in agriculture in the food industry because data from the sensors is produced from numerous locations spread across the geographical map, (Prabha et al., 2024). Farmers can also build predictive models for crop yield, pest, or the moisture level of the soil without submitting the raw data to a cloud. This not only increases privacy but also decreases the bandwidth consumption in the network. When combined with Edge AI in which an ML model is deployed and executed at grim fields such as drones, robots or IoT sensors, federated learning can enable quick decision-making that is situational at the field. For instance, an edge device may have a computer vision application that gives an almost immediate prediction of ripeness of fruits or the level of pest infestation without having to relay information to the cloud. This can significantly enhance the response time and the dependence on stable internet connections which may be an issue in the rural or agricultural regions.

Explainable AI (XAI) for Transparent Decision-Making

In general, as AI is integrated into more sensitive areas like food safety, quality assurance, and compliance, there is a heightened need for XAI. XAI is the process by which the decision-making of an ML model can be explained and understood by human beings. This is important in situations where the results of the automated decision-making process are to be relied upon by stakeholders, for instance, in rejecting contaminated batches of food or altering nutritional values. Although deep learning models are very effective, especially for highly accurate classification, they are opaque. SHAP (SHapley Additive exPlanations), LIME (Local Interpretable Model-agnostic Explanations), and Grad-CAM (Gradient-weighted Class Activation Mapping) are some of the tools that explain how the features of the input affect the output of the ML models (Yamini et al., 2024). For instance, a model designed to detect defects in foods can visually show the areas of an image that helped in the classification for the human inspector to validate. XAI is also being applied in personalized nutrition and recommendation systems to meet user's need to know why a certain diet or food is recommended. The provision of transparent explanations improves the level of trust and increases users' interaction in applications that contain health-related suggestions. The future trends in XAI in the food industry will include developing domain specific interpretability frameworks, incorporating the explanations into the decision-making dashboards and the development of standardization of the interpretability techniques for regulatory compliance.

Integration with IoT, Blockchain, and Robotics

When we talk specifically about these four domains of IoT, ML, Blockchain, Robotics – we still see the advancement in the food industry as the next generation of intelligent food systems. IoT devices produce a huge amount of real-time data from sensors placed in farms, cold chain, factories, and stores. This data when passed through the process of ML helps in predictive analysis for spoilage of products, inventory management, energy usage and the maintenance of machines. While on the other hand, blockchain technology will guarantee the aspects of data integrity, traceability, and transparency. Applying ML along with the blockchain will improve confidence in food supply chains as it will provide an immutable record of the product source, processing, and transportation history. For instance, ML models can identify temperature deviation in the records kept in blockchain, which might indicate spoilage or tampering on the way. Another area of application of robotics is also in food production and processing with the help of ML algorithms. Using vision-based ML, robots are now capable of sorting, grading and packaging according to the shape, size or quality of the products. Agrobots integrated with ML enable self-mobile devices that can identify weeds, monitor field areas, and apply pesticides in a way that is both efficient and eco-friendly. The further evolution is likely to be initiated towards closed-loop mechanisms in which the role of ML will be to facilitate constant interaction between IoT, robot assistant, and blockchain-based verification with feedback.

Sustainability-Oriented ML Applications

The role of sustainability in the food system is becoming more important in the current world, and ML is now a key enabler in the process. There are several types of applications based on the use of ML for sustainability, including food waste management systems, energy managing systems, Farm to Fork Produce, green packaging, and many others. For instance, the use of ML can improve demand forecasting through analysis of consumption patterns hence avoiding production of excessive food leading to wastage. In supply chains, ML can identify the best routes and the most efficient cold-chain operations that will reduce fuel consumption and carbon footprint. On farms, they predict the application of only the right quantities of water, fertilizers and pesticides that help save costs and also have least impact on the environment. Further, it is applied also to measure the environmental impact of food products based on the data obtained from the life cycle assessment (LCA). This makes it easier for the consumers and business entities to make better decisions on the sustainability of the available foods. As for the further development, there will be more combined optimization approaches that take into account not only the profit-making aspect but also the environmental and social ones as well. Studying the ways to incorporate sustainability indices into the loss functions of ML models will create new opportunities for ethical and environmentally sound advancements in the food industry.

Transfer Learning and Multimodal ML

Transfer Learning has become a popular technique for addressing the issue of data shortage in ML since it frequently affects food industry domains that have limited annotated data. By using large models developed on vast general databases, practitioners can achieve better local performances in food-related tasks with scarce local databases. For instance, a vision model trained on the ImageNet can be used for defect identification in grains or for plant diseases diagnosis. Also, Multimodal Machine Learning is emerging for handling/managing data from various modalities like image, text, sensors, time series data etc., in a unique model. This is especially true in the food industry where data can be obtained from various sources: the physical appearance of food products, textual descriptions of food products, values of sensors from the supply chain, and comments from the consumers. Thus, MM helps to create more complex and context-based applications, for example, image with temperature for spoilage prediction or recipe text with nutrition for meal planning. These models can also help in augmented reality (AR) tools in smart kitchens where voice commands, images and nutritional guidelines are processed at the same time. Some of the future directions in this space are as follows: Pipelining of multimodal data for a specific food industry application, developing suitable model architectures for a particular food industry application, and effective management of missing or asynchronous data in the different modalities.

CONCLUSION

Machine Learning (ML) is revolutionizing the food industry by providing solution to various problems in agriculture, food processing, safety, logistics and nutrition. This survey has demonstrated that ML is applied in various fields and areas, and how several models like neural network, decision trees, reinforcement learning and others are being used with images, sensors, text and time series data related to food. Although much has been achieved by adopting ML in improving production, quality assurance and decision-making, there are significant problems that persist. These are issues to do with data privacy, ethical issues concerning the use of personal information, interpretability of the models especially in deep learning, scalability of real-time systems, and legal framework regarding AI-based systems. To address these issues, the authors are encouraged to focus on the explainable, transparent, and resource-efficient models, especially those that can work in edge devices or the federated learning system. On the part of the practitioners, they should encourage interdisciplinary cooperation between data scientists, food technologists, and regulatory authorities as well as guarantee the validation of the ML systems in various geographical and operational settings. Therefore, scaling of ML with other advanced technologies like IoT, blockchain, and robotics can further help in refining the concept of traceability and automation along with making it sustainable in long-run. In the years that lie ahead, ML will continue to become critical in delivering on the climate change, diminishing food waste, developing targeted healthcare, and improving supply chains. Further advancements have to be planned and executed not only from the aspect of technology but also with reason and moral compass focusing on the principles of openness, and equal rights for all species as well as for the future generations. In this way, ML will continue to grow and revolutionize the food industry with its intelligent, inclusive, and sustainable solutions in the future.

REFERENCES

Anusha, S., Yamini, B., Udhayashankar, S., Priya, V., Ezhilvendan, M., Vinoth, R., & Subramanian, R. S. (2025). AI and Applied Microbiology in Sustainable Waste Recycling and the Circular Economy. In *AI Technologies for Enhancing Recycling Processes* (pp. 83–100). IGI Global Scientific Publishing.

Bhatt, S. (2024). A Comprehensive Analysis of Literature and Strategy for Eliminating Food Waste in Food Service Industry. *Promoting Sustainable Gastronomy Tourism and Community Development*, 201-213.

Boopathi, S. (2024). Advancements in machine learning and AI for intelligent systems in drone applications for smart city developments. In *Futuristic e-Governance Security With Deep Learning Applications* (pp. 15–45). IGI Global Scientific Publishing.

Gracious, L. A., Jasmine, R. M., Pooja, E., Anish, T. P., Johncy, G., & Subramanian, R. S. (2023, October). Machine Learning and Deep Learning Transforming Healthcare: An Extensive Exploration of Applications, Algorithms, and Prospects. In *2023 4th IEEE Global Conference for Advancement in Technology (GCAT)* (pp. 1-6). IEEE.

Nithya, T., Kumar, V. N., & Deepa, S., CM, V., & Subramanian, R. S. (2023, August). A comprehensive survey of machine learning: Advancements, applications, and challenges. In *2023 Second International Conference on Augmented Intelligence and Sustainable Systems (ICAISS)* (pp. 354-361). IEEE.

Oluwaseun, F. M., Bolarinwa, E. J., & Nnanna, O. C. (2024). Smart Farming [IJAEM]. *Ai Technology for Farm Monitoring and Security in Nigeria International Journal of Advances in Engineering and Management*, *6*(08), 771–781.

Pajila, P. B., Sudha, K., Selvi, D. K., Kumar, V. N., Gayathri, S., & Subramanian, R. S. (2023, July). A survey on natural language processing and its applications. In *2023 4th International Conference on Electronics and Sustainable Communication Systems (ICESC)* (pp. 996-1001). IEEE.

Prabha, D., Subramanian, R. S., Dinesh, M. G., & Girija, P. (2024). Sustainable Farming Through AI-Enabled Precision Agriculture. In *Artificial Intelligence for Precision Agriculture* (pp. 159–182). Auerbach Publications.

Prasad, M. S. C., Dhanalakshmi, M., Mohan, M., Somasundaram, B., Valarmathi, R., & Boopathi, S. (2024). Machine Learning-Integrated Sustainable Engineering and Energy Systems: Innovations at the Nexus. In *Harnessing High-Performance Computing and AI for Environmental Sustainability* (pp. 74-98). IGI Global.

Rashed, M., Fakhry, S., & Satour, R. (2025). Revolutionizing Food Quality With Machine Vision and Machine Learning Techniques. In *Food in the Metaverse and Web 3.0 Era: Intersecting Food, Technology, and Culture* (pp. 71-124). IGI Global Scientific Publishing.

Saravanan, K., Prakash, R. B., Balakrishnan, C., Kumar, G. V. P., Subramanian, R. S., & Anita, M. (2023, December). Support Vector Machines: Unveiling the Power and Versatility of SVMs in Modern Machine Learning. In *2023 3rd International Conference on Innovative Mechanisms for Industry Applications (ICIMIA)* (pp. 680-687). IEEE.

Sreelakshmi, S., Babu, A. A., Lakshmipriya, C., Gracious, L. A., Nalini, M., & Subramanian, R. S. (2024, October). Enhancing Intrusion Detection Systems with Machine Learning. In *2024 2nd International Conference on Self Sustainable Artificial Intelligence Systems (ICSSAS)* (pp. 557-564). IEEE.

Subramanian, R. S., Yamini, B., Sudha, K., & Sivakumar, S. (2024). Ensemble-based deep learning techniques for customer churn prediction model. *Kybernetes*.

Yamini, B., Prasanna, V., Ambhika, C., Anuradha, M., Maheswari, B., Subramanian, S. R., & Nalini, M. (2023). A comprehensive survey of deep learning: Advancements, applications, and challenges. *International Journal on Recent and Innovation Trends in Computing and Communication, 11*, 445–453.

Yamini, B., Ramana, K. V., Nalini, M., Devi, D. C., Maheswari, B., & Siva, S. R. (2024). Customer churn prediction model in enterprises using machine learning. *International Journal of Advanced Technology and Engineering Exploration*, *11*(110), 94.

Chapter 4
Harnessing Deep Learning for a Smarter and Sustainable Food Industry

A. N. Vinodhini
Erode Sengunthar Engineering College, India

B. Yamini
https://orcid.org/0000-0003-3531-108X
SRM Institute of Science and Technology, India

A. Revathi
https://orcid.org/0009-0003-1640-777X
R.M.K. College of Engineering and Technology, India

R. Kennady
https://orcid.org/0000-0003-0887-5007
Rajalakshmi Institute of Technology, India

T. Thilagam
https://orcid.org/0000-0002-7657-0726
Vel Tech Rangarajan Dr. Sagunthala R&D Institute of Science and Technology, India

B. Maheswari
https://orcid.org/0000-0001-9711-7915
R.M.K. Engineering College, India

M. G. Dinesh
https://orcid.org/0009-0009-7643-5338
EASA College of Engineering and Technology, India

DOI: 10.4018/979-8-3693-9884-5.ch004

ABSTRACT

DL is slowly revolutionizing the modern food sector by allowing for smart automation, quality assurance, and consumer-centric services in the food supply chain. This chapter provides a detailed description of various DL applications in the fields of food inspection, recognition, spoilage, agriculture, and nutrition recommendations. Disparate concepts of DL architectures are also discussed as well as the recent popular trends such as federated and multimodal learning and its integration with IoT and edge computing. The chapter also describes the issues concerning the quality of data, its interpretability, deployment, and the ethical issue. Finally, we provide insights on research themes that would help in the development of food systems that are sustainable, protective of the consumer's privacy, and efficient. This chapter explores how DL can be used to power innovation and resilience in the food system that is rapidly becoming data-oriented.

INTRODUCTION

Motivation: Why the Food Industry Needs Deep Learning

The food industry is one of the biggest and most significant sectors in the contemporary global economy, and it has long been experiencing the need to transform and integrate with the help of innovation. The production of food, preservation, transportation, and sale of food products that get to the consumers' tables requires more efficient, precise, and timely processes. These are as follows: food safety; reduction of food waste; and better traceability; and the need to feed the growing population with more personalized and healthier meals. The traditional methods, which are based on the manual intervention or simple rules, do not suffice for the current and future food systems. Deep learning also known as neural networks (NNs), machine learning (ML) or artificial intelligence (AI), is the newly advanced model to handle various kind of data like image, time series signal and text data, (Kumar & Dwivedi, 2022; Sudha et al., 2023). DL models are able to learn complex patterns and representations from big data sets without the programmer's intervention. This makes them particularly useful in the field of food industry as the data is gathered from numerous sources, including visual observations, sensors, records of sales, and customers feedback etc., (Kashyap, 2019). For instance, DL models have enhanced food quality inspection by analyzing images of the produce or packaged food items, which is more effective than computer vision methods. They are also being employed in checking on food spoilage, shelf-life estimation, nutrition recommendations and even transportation management. Considering the variety of these applications and the growth of digitalization in the food industry, the need for the use of deep learning in this sphere is higher than ever, (Chinnasamy et al., 2023).

Objectives and Scope of the Survey

This survey is aimed at investigating prior and existing studies and real-world applications of deep learning in the food industry extensively. Thus, the chapter is designed to cover as many aspects and stages of the Food life cycle as possible, as well as to provide a deeper understanding of how DL is used in each of them. This survey is therefore aimed at a wide cross-section of the population that includes researchers, data scientists, food technologists, agronomists, supply chain managers and policymakers. For all professionals, it provides an opportunity to see new areas of application of artificial intelligence and real-life problems that need to be solved with the help of efficient and scalable AI methods. For the practitioners in the food industry, it presents a real-life application of how the deep learning could be adopted to open up new opportunities and address current challenges. This survey is not restricted to a specific area of application but rather it covers a vast area of concern. It comprises food image classification, food quality and safety assessment, crop inspection through remote sensing, demand prediction, and dietary recommendation. Furthermore, the chapter also covers interdisciplinary topics like the integration of DL with IoT, edge computing, and federated learning which can be useful in food industry to deal with specific issues like data privacy and real-time decision making.

Contributions of this Chapter

Thus, this chapter has several valuable contributions to the research about deep learning and food industry-related topics.

Scoping Study

It provides a broad identification and evaluation of DL applications in the food industry including yield prediction of agricultural crops and customized meal planning. All the areas of application are described in relation to the type of DL architecture applied, the type of data required, and real-life application.

Assessment of Model Characteristics

The chapter also describes which types of deep learning models (CNNs, RNNs, LSTMs, GANs, and Transformers) are appropriate or unfit for solving certain problems based on the nature of these models.

Identification of Key Challenges

It discusses the issues that are both functional and technical when it comes to implementing deep learning in food-related scenarios. Some of these are problems such as lack of data, interpretability of the models, limited computational power and real-time analysis.

Although the current trends are indicated in the survey, they also embrace new trends and future of AI including multimodal learning that deals with fusing image, text and sensor data, Artificial Intelligence that does not infringe on privacy through Federated learning and sustainable and Ethical AI.

Structure of the Survey

This chapter is divided into several subheadings that capture the different aspects of DL in the food industry.

Section 2

Background and Fundamentals of deep learning provides the context needed to understand deep learning and covers a few key architectures as well as tools that can be used. It also outlines the goals for the food industry in the digital transformation, which would be useful for readers who are not from an AI background.

Section 3

Applications of Deep Learning in the Food Industry is the main part of the chapter that will focus on the detailed description of DL applications in the food industry. These are the quality inspection, food categorization, spoilage forecast, precision farming, demand chain optimization, nutrition personalization, and food security.

Section 4

Key Challenges explains that the major difficulties of implementation include the absence of numerous datasets, high deployment costs, and ethical issues. It also discusses the technical drawbacks of the models under consideration and indicates the directions of further research. The fifth section entitled Emerging Trends and Future Directions provides the reader with an outlook of the future of the field. It further discusses the integration with corresponding technologies, the use of unsupervised and few-shot learning, and the growing emphasis on sustainability an inclusion into the AI Food Systems. Section 6: Conclusion is the last part that restates the findings made in the chapter and presents an outlook on further research.

BACKGROUND AND FUNDAMENTALS

Overview of Deep Learning

Deep learning (DL) is a subfield of AI that deals with deep neural networks with multiple layers as presented in the name, (Prabha et al., 2022) These models mimic the functionality of the human brain and follow the approach of generating feature representations on their own from the data with suitable architecture as to cater for problems like image recognition, understanding of natural language processing and sequential decision making giving a high precision rate, (Venkateswaran et al., 2023). Deep learning has emerged as an effective tool in the field of machine learning due to the better results it provides compared to the conventional methods in the large data set, (Anusha et al., 2024).

CNN is one of the most famous architectures in the deep learning field and has significantly changed the image processing tasks. CNNs are especially suitable for inspecting food quality, recognizing the food and validating the packaging since they are proficient in Unraveling spatial hierarchies of image data, (Ganesan et al., 2024). By constructing multiple layers of convolutions, pooling as well as activation functions, CNNs are capable of extracting high-level features that are required for food type classification or defect detection in products, (Kumar et al., 2023).

Figure 1. Deep learning in the food industry

The next significant type is Recurrent Neural Networks (RNNs) intended for data of sequences. RNNs have the ability to hold some internal states making them a good fit for time series data such as temperatures in a food storage, sensors in logistics chains, or customers' behavior over the time. The LSTM network that is a more developed version of RNNs resolves the vanishing gradient problem and allows the model to remember the long-term dependencies, (NP et al., 2024). LSTMs have been used in predicting shelf-life and tracking the spoilage process using previous sensor or environmental information.

Another type of deep learning models is Generative Adversarial Networks (GANs) that include two neural networks: generator and discriminator. It is noteworthy to mention that GANs are powerful enough to generate synthetic data that seems real. In the food industry, GANs have been used in data augmentation including in producing new realistic images of food that can be used to train better classifiers particularly where there are limited labeled datasets. They also have an application in generating synthetic sensor data and enhancing the reliability of the models applied to safety supervision.

Recently, Transformers have become quite a popular architecture, which were initially used in the NLP field but are now adapting to computer vision and time-series forecasting. Self-attention mechanisms are used to assign weights to input elements, which makes transformers capture dependencies. In the food domain, Transformers can be used for: translating the list of ingredients, for analyzing the reviews, or for processing multiple inputs as images, text, and sensors.

For tools and frameworks, deep learning is indeed poorly supplied but this necessity has led to their development and now users can find a lot of them. Python frameworks such as TensorFlow, PyTorch and Keras offer friendly API, whether for model construction, training or deployment. These frameworks support GPU acceleration which makes it easier to handle large data set which is common in most of the food applications. Other tools that are used in the development of DL include OpenCV for image processing, Scikit-learn for preprocessing, and ONNX for the deployment of the developed models in different platforms.

Deep learning models have high flexibility and scalability that makes them suitable when it comes to AI application in the food industry. The increase in availability of data from agricultural drones, food processing lines, point of sale systems, and consumer applications makes deep learning the best way of making use of the raw data.

Overview of the Food Industry's Digital Needs

The world of food industry is rapidly changing due to the impact of digitalization. This is especially significant because global supply chains are becoming increasingly intricate and customer demands continue to rise for organizations within agribusinesses, manufacturing, distribution and retail industries. Deep learning as one of the key approaches to modern artificial intelligence is in synergy with these digital initiatives and provides solutions to some of the most acute problems of the industry.

The three major requirements for the current food industry include the following: Automation. From picking fruits and sorting them mechanically to quality inspection in industries, companies wish to reduce the amount of manual effort while increasing the rate of production and ensuring quality. This is where deep learning comes in handy as it allows the machines to have a vision of the world around them. For example, using visual inspection systems based on CNN it is possible to detect bruises on fruits, illegible labels on packages or other foreign objects in the production line more accurately than in the case of the application of traditional rule-based systems.

Another important area is also the quality assurance. The population of consumers today is more so conscious of their health and the brands they are interacting with, thus their demands for quality and standards are very high. It is in this regard that DL models assist in maintaining these standards by recognizing any variation in appearance, texture or packaging. Besides, through analyzing the historical production and environmental data, issues can be predicted, and necessary changes can be made in advance.

Nowadays, there is a growing focus on traceability and transparency because of constantly growing cases of food recalls and safety issues.

Different governments and the consumers as well have a desire to know how the food products are handled and what the food products are made of. Deep learning helps to achieve these goals by processing data at different levels of the supply chain, from the images of fields obtained from satellites to sensing data in cold stores and identifying possible risks or deviations. In addition, the NLP-based models can analyze text from the news articles, inspection reports, and social media, and generate alerts of potential contaminations or fraudulent activities.

Consumer analytics is another domain in which DL is beginning to make a large impact . Food companies now have a great opportunity to gather a lot of information on the consumers due to the increasing popularity of e-commerce and smart kitchen devices. This data can be analyzed by DL models for recommending products, identifying trends of consuming specific food items, and even the sentiments of the public towards a new product.

For instance, transformer based language models can analyze the reviews and comments to determine a trend while the neural collaborative filtering can be used in creating a nutrition platform.

The role of data in the production and supply of food cannot be overemphasized in the modern world. Whether it is predicting the demand, managing inventory or planning the delivery routes, the business needs to make accurate predictions to avoid wastage. LSTMs and CNN-RNN models are particularly effective at predicting the sales based on historical data, weather information, and market trends. These forecasts in turn assist in avoiding overproduction, perishable stock, and better matching of supply against demand.

In addition, the use of edge computing and IoT in food environments implies that data is being captured in real time from smart scale, temperature sensors, and packaging monitors. The data collected can then be processed using deep learning models at the edge and make real-time decisions such as determining when a refrigerator is faulty or when a food item has been left out of refrigerator for a long time. This capability is important in preserving food and reducing cases of food wastage, (Balakrishnan et al., 2024).

In other words, the five areas that are most important to the needs of the food industry are automation, quality, traceability, customer relations, and optimization, (Jebamalar et al., 2023). Deep Learning in a way is an enabler for all of these areas, providing not only accelerating but also intelligent solutions to address the problems presented. Thus, it can be assumed that the interaction between deep learning and the food sector will continue to strengthen as the latter becomes more integrated into global and digital processes, which will lead to the development of new products and services for consumers.

APPLICATIONS OF DEEP LEARNING IN THE FOOD INDUSTRY

Food Quality Inspection and Grading

Another area that is widely implemented in food industry is using deep learning for the visual inspection and grading. In the past, the process is done manually, and it is time-consuming, non-repeatable, and can be inaccurate. Convolutional Neural Networks (CNNs) are part of the deep learning option that can be used to automatically analyze the food items and their packaging.

CNNs have been widely used in detecting surface defects of fruits, vegetables, and grains. They are also able to identify slight variations in texture, color and shape which are the signs of ripeness, bruising or contamination respectively. Deep learning models have been used in various aspects such as marbling, fat distribution and quality grading of meat. In a similar way, in the packaged goods, CNNs are applied in label misalignment, broken seals, and foreign particles. The application of these models in the industry is usually accompanied by the use of high-speed cameras and conveyors. Real-time image acquisition then followed by the result from the trained CNNs allows for quick sorting and decision making. They also minimize wastage and offer a better quality while at the same time meeting food safety standards.

Figure 2. Applications of DL in the food industry

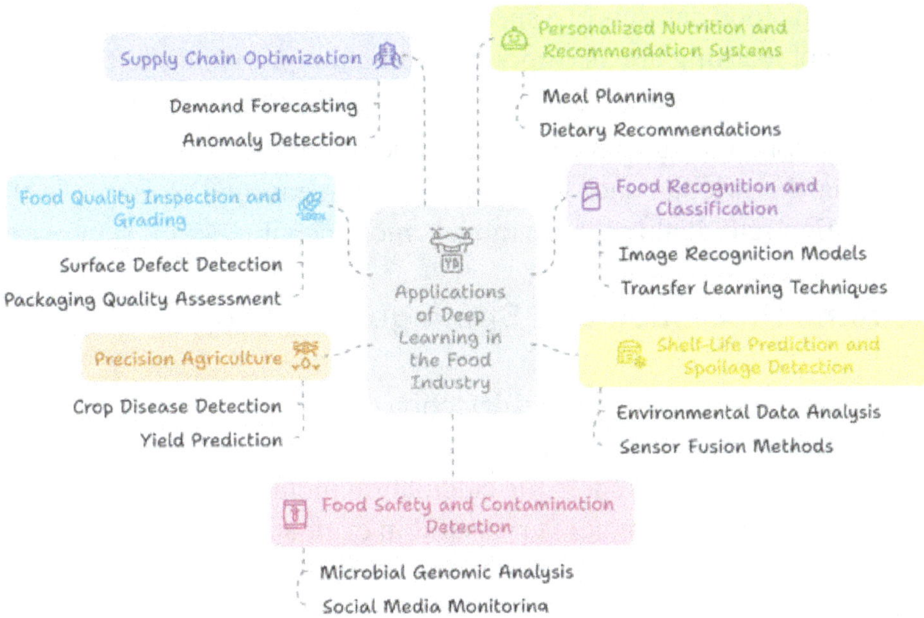

Food Recognition and Classification

In the post, Deep learning has also contributed immensely to food recognition and even classification. It is most commonly used in the MHIS, smart kitchens, and retail stores. The CNNs and pre-trained image recognition models like ResNet, Inception and MobileNet can distinguish between the food items from images with high precision, even when the categories of food items are similar in terms of their appearance.

In consumer-oriented applications such as calorie intake and diet tracking, the food recognition systems use uploaded images of food to determine what the food is and its portion size. In self-checkout systems, similar models replace barcode scanning to ensure no physical contact is made during the purchasing process.

However, these systems have some issues like intra-class variation (for example, pasta can be of different types) and inter-class resemblance. To overcome these problems, transfer learning is employed in which a model that has been trained on a large image dataset (such as ImageNet) is further trained on the food datasets of a specific size. Data augmentation methods and few shot learning are also applied in an attempt to manage the issues arising from the limited amount of data in specific food categories.

Furthermore, the current studies consider the integration of the image and textual knowledge such as the recipe or menu text to improve the classification rate, making these models more flexible.

Shelf-Life Prediction and Spoilage Detection

Determination of shelf-life and detection of spoilage are very important in order to minimize food wastage and also to enhance food safety. The conventional methods of shelf-life calculation include the use of specific dates of expiry, which are not very useful in evaluating the actual state of the product at any given time. RNNs and LSTM networks in general are a more dynamic and accurate approach to the problem.

The environmental data such as temperature, humidity and gas levels can be learned with LSTM networks due to their capability to capture the long-term dependencies in the time series data. For instance, machine learning models that have been trained on historical data of cold storage can be used to forecast spoilage patterns and consequently come up with ways of improving the storage conditions, (Prabha et al., 2022).

Apart from time series data, spoilage detection also uses sensor fusion methods. All information gathered from the visual system, olfactory sensors or electronic nose, and chemical sensors are used by deep learning models to determine spoilage. CNNs are applied in the detection of changes in the appearance of fresh produce or meat, while the multimodal networks handle these inputs as a whole.

These DL models can be implemented in IoT systems that monitor the food conditions and raise the alarm when the spoilage risk is identified. This is a proactive approach that helps to increase the shelf life of the foods, minimize wastage and enhance consumer health.

Precision Agriculture

That is, that precision farming brings an idea to optimize farm production and efficiency using information-aided technologies. In this context, deep learning is a powerful tool that helps analyze all the information related to agriculture without human intervention, (Prabha et al., 2024).

Crop disease detection is one of the significant applications of the proposed model. The CNNs can detect visual signs of plant diseases at an early stage either by using drone or smartphone imagery, (Rajendran et al., 2023). These models are faster and more accurate than the regular approaches and can be used in the field through mobile applications or be incorporated in UAV platforms. There are usually pixel-wise segmentation models such as U-Net and DeepLab which can be used to identify the affected crop areas for treatment.

Another critical application is yield prediction which entails predicting crop yield arising from factors they include weather conditions, soil and health detail of plants. LSTM networks and other models where CNNs are combined with RNNs have been used in yield forecasting based on satellite imagery and sensor data over time.

In addition, satellite and drone imagery with DL models help in the identification of weeds, count the plants and optimize the use of inputs by farmers. These capabilities help in enhancing yield, reducing cost and being environmentally friendly.

Supply Chain Optimization

The food supply chain is the process of transferring food through several stages from producers all the way to the end consumers. These operations are, therefore, made efficient by the deep learning models by way of forecasting, detecting anomalies, and decision-making, (Siva Subramanian & Prabha, 2021).

The application of demand forecasting is among the most important ones. LSTM models may be used to forecast future sales by using the previous sales data and considering the factors like the season, weather, and the campaigns. This results in better planning of procurement, production and distribution activities, hence ensuring that the organisation achieves its goals and objectives.

In logistics, the anomaly detection models are employed to point at the disruptions or inefficiencies. These are CNNs and autoencoders that seek to identify unusual patterns in delivery routes, temperature records, and movements of products in stock so as to detect cases of theft or spoilage or delayed delivery.

There are also some studies done in the field of RL for inventory and warehouse management. RL agents are trained to make the stocking and routing decisions in the simulated environments that has the minimum cost and maximum freshness. Therefore, by applying DL for supply chain management, companies can minimize costs, avoid delays, and adapt to the market conditions effectively.

Personalized Nutrition and Recommendation Systems

There is an increasing concern in today's society regarding health and fitness, which has led to an increased focus on nutrient individualization. Deep learning improves recommendation systems for recommending meals and dietary plan based on one's preference, medical condition and desired outcomes.

The DL-based systems employed in the context of this research include collaborative filtering, NLP, and image analysis to establish the user's behavior. For instance, logging and monitoring applications that involve food image recognition through CNNs record consumed meals, and model Transformer performs the assessment of logs and food preferences and suggests better options.

Personalized recommendation engines involve personal health information including age, level of activity, allergies and diseases among them. Based on this information, the DL models then create meal plans or grocery lists that are acceptable from nutritional standpoints. Neural collaborative filtering and autoencoder are other methods of creating recommender systems that enhance the above suggestions through feedback.

Some of the more complicated use downstream knowledge graphs and multi-modal learning – which include database of nutrition, User engagement, recipe contents and even live biometrics from the user-wearable. These systems do not only enhance learners' participation but also enhance healthy lifestyles as recommendations on diet are made easily understandable in these systems.

Food Safety and Contamination Detection

Managing food safety is a major issue of public health, and deep learning is an effective approach to preventing and avoiding contamination. Based on microbiological, chemical and environmental information, it is possible to predict the likelihood of contamination at the production, transportation and storage stages.

For instance, the DL models can be used to study the genomic sequences of microbes and anticipate their actions; thus, in food hygiene, the models can be used to detect bacterial intrusion in ready-to-eat meals. CNNs have been applied in analyzing microscopy or sensor data to identify pathogenic organisms in foods.

NLP on the other hand complements by listening to unstructured text sources. Some of the transformer-based models can be used to monitor social media, news, and regulatory databases for signs of outbreak, recall or any complaints. This helps companies prevent harm or initiate a recall reducing the risk and the potential harm posed to the general public.

DL also responds to fresh food product packaging systems with sensors able to monitor freshness indicators conveying data concerning spoilage. Integrating with edge computing means that the inference can be performed with low latency and immediate alerts in distribution networks.

In a more complex way, deep learning is improving the efficiency of risk management in the food industry and strengthening its ability to meet the current and future requirements, including higher levels of regulation.

KEY CHALLENGES

However, there are several issues which are worth discussing while analysing the further development of deep learning (DL) application in the food industry. These are technical, operational, and ethical in nature and solving them is crucial for broad and long-term adaptation. Here, the main challenges that are still open for DL in the food domain are discussed, including data issues, interpretability, deployment issues, and legal aspects.

Data Quality and Availability

Deep learning depends on big and various and high-quality data sets. However, in the food industry, such information is still hard to come by. It is unlike sectors such as healthcare or finance where data is relatively clean, less noisy, and available in abundance at times in the food sector, data may be limited in certain categories especially those that are small or specialized. For instance, training a good food classification model needs several thousands of images per class, however, traditional or homemade foods are scarce in public datasets.

Also, the data collected from the farm, factory, store, or consumers is different in terms of structure and credibility. The sensors may not always provide complete and accurate information, the images may be obstructed or poorly illuminated, and the metadata may not be correct or even missing. These quality issues have a direct relation to the effectiveness and reproducibility of DL models.

However, there are also some factors that hinder data sharing among the organizations; these are privacy and proprietary. The main reason why food companies do not engage in information sharing with other companies or organizations is that they have legal or competitive constraints on the type of information they release to the public. It results in having isolated datasets and does not help in the creation of more generalizable models.

In order to mitigate these challenges, data augmentation, synthetic data generation (e.g., GANs), transfer learning, semi-supervised learning have been adopted. But they fall short of solving the problem only in the partial sense. deeper with the science and engineering developments that are less likely to be achievable in the long run without the open-access source datasets, with new collaborations of academics and industries, and health standards for protecting privacy while sharing data (federated learning).

Interpretability and Trustworthiness of DL Models

Another limitation of the use of deep learning is that it is not easily interpretable or explainable. While most DL architectures including deep CNNs and Transformers are 'black-boxes' that make decisions based on high-level abstract features, the proposed system generates intermediate feature maps that are easier to interpret by humans.

In the food industry where its choices could have consequences on the health of the population, trust by the customers, and the law this gives a lot of concern. For instance, a DL model employed to identify contaminated foods or recommend a diet plan must be able to explain its decisions to the food scientists, regulatory bodies, and the consumers.

This is especially compounded by the fact that false positives and false negatives are often disastrous in safety-sensitive applications such as spoilage identification, microbial threat identification, or allergen alerts. Thus, in such cases, the stakeholders require not only high accuracy but also XAI approaches that reveal how the model arrives at the decision.

There are several approaches that are being developed to improve the interpretability. These are Saliency maps, Grad-CAM, LIME and SHAP. They also assist in identifying which areas of the image or input influenced a decision to a large extent. In models based on time series, it is possible to see which particular time steps are more essential or which features have the most relevance.

Althogh 'Interpretability tools' are emerging nowadays, they are yet to be fully refined and optimized especially when applied to food-related functions. It is important to have a closer cooperation between AI researchers and food domain specialists to provide both technically sound but also socially relevant explanations.

Real-Time Constraints and Deployment Issues

A large number of deep learning applications in the food industry are real-time or nearly real-time in nature. For example, quality inspection in a production line, disease identification in cold supply chain, or check-out systems in supermarkets cannot tolerate high latency. Nonetheless, applying DL models in such environments has some challenges as discussed below.

First, it should be understood that increasing the number of parameters and factors affects the feasibility of real-time computations. Transformers and deep CNNs are very demanding in terms of computational resources and memory. Although the cloud computing grants the concept of elasticity, concerns regarding latency and connectivity challenges make it a non-starter when it comes to edge use cases such as drones or a manufacturing floor or a mobile device, (Dhinakaran et al., 2025).

Second, creating a real world map requires stability and versatility. Models trained in ideal conditions without interference of factors such as lighting, different hardware, or other food types perform poorly when exposed to such conditions. This domain shift leads to decrease in performance, which requires retraining or adaption all the time.

Third, it is important to notice that integration with the legacy systems is not an easy task. Most food processing facilities are old and integrating deep learning systems in the existing production lines entail both technical compatibility and considerable capital investment.

To overcome these issues, current researcher proposals involve model compression methods such as pruning or quantization, optimized network structures like MobileNet or EfficientNet or efficient deployment models of artificial intelligence at the edge. Further, model retraining pipelines and continual learning strategies are being explored in order to achieve the capability of the model to adapt to dynamic environments.

Real time is still a difficult problem in practice and it is not a simple task to design a real time system that is fast, accurate and does not consume too many resources.

Regulatory and Ethical Concerns

It is critically important to discuss the role of deep learning in the current informatics and management of food systems, as the problems of both the regulation and ethics of this technology are emerging. The governments and international organizations are gradually developing rules and regulation for the application of AI in high-risk sectors such as food.

The areas of primary focus with regard to regulation include compliance and accountability. If a DL model fails for example in its ability to classify a contaminated product or in giving out wrong nutritional information then who is to blame? These concerns make companies avoid relying on the decisions made by AI. In addition, there may be legal obligations for the AI systems to produce logs of the decision making process or justification for certain decision.

Data privacy is another important area of concern which may play a large part in personalized nutrition and consumer metrics. DL systems get and process a lot of sensitive user data such as the user's eating habits, personal health, and location.

Other issues are related to ethical concerns such as the issue of bias in the algorithm and the issue of fairness. The models trained on such pre-distributions of data may have a poor performance for certain minority categories of people. For instance, a calorie estimation model learnt and developed on Western foods may fail to recognize or completely disregard traditional foods from other parts of the world. This can culminate into instances where such systems enhance dietary prejudices and segregation.

To handle these questions, it is necessary to work with ethical guidelines for AI that involve openness, non-discrimination, and accountability. Some of the approaches used modern proactive actions and measures include bias auditing, data diversification, and human-in-the-loop systems.

At last, it is worthy to note that the environmental effects of deep learning are also significant. Training large DL models is energy-intensive that make it add to the carbon ft. Since sustainability has become a global concern, especially in the food industry, scientists have sought green AI, which are accurate AI models that are also energy-efficient.

Therefore, despite the opportunities that deep learning provides for the food industry, several crucial issues should be taken into consideration. This chapter discusses that data quality, model interpretability, real-time deployment, and compliance with the existing regulation are critical for the proper use of the AI models. With the changes undergone in the AI food field, cross-functional cooperation between Computational intelligence scholars, food scientists, ethicists as well as policymakers is to be instrumental in establishing a world where deep learning positively impacts the safety, intelligence as well as sustainability of food production.

EMERGING TRENDS AND FUTURE DIRECTIONS

Indeed, with the increased development of DL, the possibility of its application in the food industry seems more realistic. However, to fully exploit this potential, new directions are developing that may help to avoid the existing problems and open the new opportunities. These emerging trends are as follows: technological with IoT and edge computing, Multimodal, Privacy-Preserving Learning, Sustainability and Generalization, (Sudha et al., 2023). This section elaborates these transformational trends and points to how they are likely to define the future of future AI-based alimentary production system.

Integration with IoT and Edge Computing

Another interesting area of development in the use of deep learning in the food industry is the combination of DL models with the IoT and edge computing. With the more food processing facilities, farms and retail outlets being fitted with sensors; temperature, humidity, footage of produce or even consumers, real-time data is produced in large quantities.

Figure 3. Emerging trends and future directions

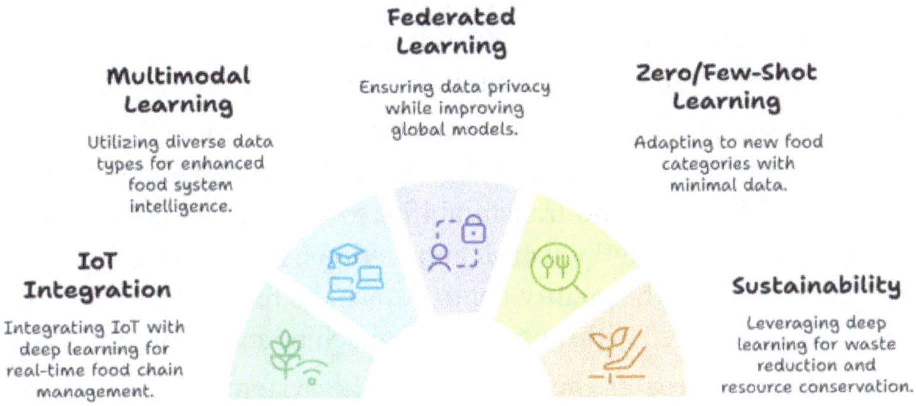

Through the end-vehicle deep learning models, real time inference is made possible without high latency than cloud-based system in devices as drones, smart cameras and embedded processors, (Ponnuviji et al., 2025). For example, in precision agriculture, for instance, the lightweight CNN models can be carried in drones whereby the drones detect crop stress or diseases during their flight. In smart refrigerators or warehouses, DL-based sensors can track the shelf-life indicators and notify when the products are near their expiration dates. Combination of DL and IoT is paving way for an intelligent food chain where monitoring, predicting and controlling the food chain is made possible at real time, (Sudha et al., 2024).

Use of Multimodal Learning

Since food related activities are not simple, it is quite natural that some tasks involve the use of different types of data. For example, recognizing a food item may require one to identify the item through an image, a written or spoken description and even the nutrients value of the food. Likewise, knowledge of spoilage might involve the use of visual signals, gaseous signals, and temperature records. In order to address such situations, multimodal deep learning is emerging as a popular solution.

Multimodal learning is a learning paradigm where models are learned to work together with multiple forms of data such as images, text, audio and sensor inputs. In the context of the food industry, this means creating models that can take in images of food, text data in the form of recipes and food labels, tabular data containing nutrients and sensory data.

Recent Researches on Vision Transformer Models as CLIP and Vision-Language Models suggest that the alignment of both vision and language yields better robustness and transferability to the classification and recommendation systems. In industrial applications, multimodal models are employed to improve the quality control since the results from the vision system can be compared with those from the machine and human inspection.

It is not only increases the accuracy of the system but also makes it context-aware. For instance, an application that is related to food can recommend meals that the user may like to take depending on his or her past food intake, the food that the user has at home through barcode scan or photo and the level of physical activity the user is likely to undertake as identified by the fitness trackers.

Multimodal deep learning is expected to have a significant impact on further development of DL based food systems as intelligent, flexible and user-friendly.

Federated Learning for Data Privacy

One of the major challenges that hinder the use of DL in the food sector is sensitivity and decentralization of data. Sometimes, it is difficult for companies to provide consumer data, secret recipes, or even internal quality records because of privacy, competition, and legal requirements.

Another solution for this problem is provided by federated learning (FL) that allows for model training without exchanging the data. FL specifically works in such a way that each organization or device uses its data and trains a model locally, but only the weights of the model are sent to the central server for averaging. This way, data will only remain within its source while at the same time will be useful in enhancing a global model.

In the food industry, for instance, many parties, including retailers, producers, and logistics firms, can train and update a model for demand forecasting, spoilage identification, or recommender system without sharing sensitive data. It is especially helpful in personalized nutrition which involves health and diet information that can be used in improving models locally without violating the user's privacy.

FL also poses some difficulties like model synchronization, non-IID data distribution, and communication load. But, in the recent years, certain techniques like differential privacy, homomorphic encryption, and secure aggregation have made the federated learning more practical and secure.

In this way, FL creates the basis for more effective, comprehensive, and privacy-preserving DL applications in the food domain.

Zero-Shot and Few-Shot Learning in Food Applications

It was widely recognized from that deep learning is restricted by the requirement of large amounts of labeled datasets. This is especially true in the food industry where certain dishes or products may be rare, culturally sensitive, new in the market or have a unique spoilage situation where annotated data may not be available.

This problem is now being solved with the help of zero-shot and few-shot learning methods. These techniques are designed to be used when there is little or no training data in the new task or a new category. Zero-shot learning allows the system to learn new food categories using the textual or semantic description of the food items by using models trained on large-scale data.

Hence, few-shot learning is a process of fine-tuning models on only a few examples that have a label on them. As a result, meta-learning, prototypical networks, and contrastive learning are used in food classification, generating recipes, and nutrition estimation in low data settings.

These methods are particularly useful in dynamic environments, for instance, in the case of new product, a new trend in diet, or specific sectors in certain regions. They also minimize the cost and time required in constructing and implementing DL systems and therefore beneficial for startups and small food businesses.

With models getting more efficient in terms of the amount of data required for learning, the concept of zero and few-shot learning will go a long way in making artificial intelligence more accessible to the global food industry.

Sustainability and Food Waste Reduction via DL Insights

Sustainability is now one of the values that are very important in the food industry, and deep learning can play a significant role in waste management, resource utilization, and environmental control. DL models can also be used to analyze production, storage, consumption patterns in order to determine where wastes can be avoided and resources conserved. For instance, using DL based demand forecasting for retailers, they can reduce the cases of overstocking perishable products and hence reduce spoilage. Smart packaging or IoT sensors when integrated with spoilage detection systems may help in giving real-time information and therefore, real-time response in terms of pricing or redistribution of near-expiry items. In agriculture, deep learning means analysis of the soil quality, activity of pests and diseases, crop health, and provided accurate utilization of water, fertilizer and pesticide. Other models for managing cold chain ensure that food is kept at the right temperature during transportation so that it does not spoil easily thus reducing losses. In addition, it is possible to use DL models to detect patterns of consumer behavior that leads to overconsumption or waste and apply it to awareness campaigns or app-based prompts. When incorporated into recommenders these tips can enhance standards of sustainable consumption for goods such as locally available or seasonal foods.

As sustainability is shifting from being a competitive asset to the need for the survival of the planet, deep learning will not only be employed for profit purposes but to save the planet. The fourth generation of deep learning in the food industry will encompass its interaction with other technologies, its performance under conditions of limited data, the privacy and inclusiveness of the algorithms as well as its sustainability. These trends will not only extend the applicability of DL systems but also make the systems more usable, reliable and consequential. At the same time, information and ideas that enhance deep learning, along with technology, ecology, and broader innovations, will define a better, safer, and sustainable food future.

CONCLUSION

AI in food industry has become one of the most crucial technologies that have revolutionized food chain from production, processing, distribution, and retailing of food to consumption. It has numerous usages such as quality inspection for product, health monitoring for food, smart shopping, shelf-life and supply chain forecasting, and recommendation systems for people's dietary preferences. These innovations are empowering a transition from traditional style to a scientific approach, monitoring, and enhanced consumer satisfaction. However, deep learning poses some challenges when integrated into food systems as explained below. Some of the challenges include lack of adequate data, lack of standard labelling, and poor standardization that affect the model's training and generalization. Thus, some of the general issues for DL models include: interpretability and trust, as these are black-box models and can be used in safety-critical or regulated areas. However, the application of DL solutions in real-time has limitations concerning computational and infrastructures requirements, as well as privacy, security, and ethical issues. These challenges can be fully tackled through further investigate on the explainable AI, the development of novel, light, and edge models, federated learning for private cooperation, and generalizable approaches including zero- and few-shot learning. At the same time, it is necessary that the industry takes up new trends like multi-modal learning, IoT integration and sustainable analytics to optimize DL. This is the course of creating cooperative partnerships between AI researchers, food scientists, ethicists, and other representatives of such industries that would create accurate and sustainable systems. It can be hypothesized that as the field of deep learning progresses further it will become increasingly integrated in the food industry in a manner that will aid in solving problems related to food safety, reduction of avoidable wastes in the food chain, and the promotion of good human health culminating to the realization of a smarter, safer, more effective food chain.

REFERENCES

Anusha, S., Ezhilvandan, M., Gayathri, S., Sivaraman, V., & TP, A. (2024). Emerging trends in edge AI for Industry 4.0 and 5.0: Technologies, applications, and challenges. *AI Innovations in Service and Tourism Marketing*, 211-228.

Balakrishnan, C., Anish, T. P., Nithya, T., Sheena, B. G., Maheswari, B., & Subramanian, R. S. (2024). Revolutionizing Trust Manufacturing Integrating Real-Time Data and AI Innovations for Enhanced Efficiency. In Industry Applications of Thrust Manufacturing: Convergence with Real-Time Data and AI (pp. 74-97). IGI Global.

Chinnasamy, P., Sathya, K. S., Jebamani, B. J. A., Nithyasri, A., & Fowjiya, S. (2023). Deep learning: algorithms, techniques, and applications—a systematic survey. In *Deep learning research applications for natural language processing* (pp. 1–17). IGI global.

Dhinakaran, D., Kumar, N. J., & Ponnuviji, N. P. (2025). Safeguarding confidentiality and privacy in cloud-enabled healthcare systems with spectrasafe encryption and dynamic k-anonymity algorithm. *Expert Systems with Applications*, *279*, 127584.

Ganesan, I., Ponnuviji, N. P., Kumar, A. S., Nithya, M., Jambulingam, U., & Lalitha, S. D. (2024). Real-Time Event Detection and Predictive Analytics Using IoT and Deep Learning. In Industry Applications of Thrust Manufacturing: Convergence with Real-Time Data and AI (pp. 1-41). IGI Global Scientific Publishing.

Jebamalar, B., Sudha, K., Maheswari, B., Nalini, M., Dinesh, M. G., & Subramanian, R. S. (2023, May). Collaborative filtering for bargaining and customer feedback in e-commerce. In 2023 7th International Conference on Intelligent Computing and Control Systems (ICICCS) (pp. 1804-1808). IEEE.

Kashyap, R. (2019). Sensation of deep learning in image processing applications. In *Handbook of Research on Deep Learning Innovations and Trends* (pp. 72–96). IGI Global.

Kumar, A., Bhatia, A., Kashyap, A., & Kumar, M. (2023). LSTM network: a deep learning approach and applications. In *Advanced Applications of NLP and Deep Learning in Social Media Data* (pp. 130–150). IGI Global.

Kumar, R., & Dwivedi, R. K. (Eds.). (2022). *Advances in Deep Learning Applications for Smart Cities*. IGI Global.

Ponnuviji, N. P., Nirmala, G., Kokila, M. S., & Priyadharshini, S. I. (2025). Optimizing Image Retrieval in Cloud Servers with TN-AGW: A Secure and Efficient Approach. Journal of The Institution of Engineers (India): Series B, 106(2), 459-473.

Prabha, D., Maheswari, B., & Aswini, J. (2022). Wrapper-Naive Bayes approach to perform efficient customer behavior prediction. In Innovative Data Communication Technologies and Application [Singapore: Springer Nature Singapore.]. *Proceedings of ICIDCA, 2021*, 17–31.

Prabha, D., Subramanian, R. S., Dinesh, M. G., & Girija, P. (2024). Sustainable Farming Through AI-Enabled Precision Agriculture. In Artificial Intelligence for Precision Agriculture (pp. 159-182). Auerbach Publications.

Rajendran, T., Ramana, T. V., Nalini, M., Subramanian, R. S., Devi, D. C., & Anuradha, M. (2023, July). Paddy Plant Leaf Disease Classification based on Artificial Intelligence. In 2023 4th International Conference on Electronics and Sustainable Communication Systems (ICESC) (pp. 1002-1009). IEEE.

R.S. Sudha, K., Ramana, K. V., SivaKumar, S., Nithyanandhan, R., & Nalini, M. (2023, February). Hybrid Variable Selection Approach to Analyse High Dimensional Dataset. In 2023 7th International Conference on Computing Methodologies and Communication (ICCMC) (pp. 1489-1495). IEEE.

Siva Subramanian, R., & Prabha, D. (2021). Optimizıng Naive Bayes probability estimation in customer analysis using hybrid variable selection. In Computer Networks and Inventive Communication Technologies: Proceedings of Third ICCNCT 2020 (pp. 595-612). Springer Singapore.

Sudha, K., Ambhika, C., Maheswari, B., Girija, P., & Nalini, M. (2023). AI and IoT applications in medical domain enhancing healthcare through technology integration. In *AI and IoT-Based Technologies for Precision Medicine* (pp. 280–294). IGI Global.

Sudha, K., Balakrishnan, C., Anish, T. P., Nithya, T., Yamini, B., Subramanian, R. S., & Nalini, M. (2024). Data Insight Unveiled: Navigating Critical Approaches and Challenges in Diverse Domains Through Advanced Data Analysis. Critical Approaches to Data Engineering Systems and Analysis, 90-114.

Venkateswaran, N., Vidhya, R., Naik, D. A., Raj, T. M., Munjal, N., & Boopathi, S. (2023). Study on Sentence and Question Formation Using Deep Learning Techniques. In *Digital Natives as a Disruptive Force in Asian Businesses and Societies* (pp. 252–273). IGI Global Scientific Publishing.

Chapter 5
Optimization Algorithms in Food Product Quality Assessment

Usharani Bhimavarapu
https://orcid.org/0000-0002-0246-1420

Department of Computer Science and Engineering, Koneru Lakshmaiah Education Foundation, Vaddeswaram, India

ABSTRACT

Improving the food industry requires innovative techniques for quality assessment, particularly in evaluating the characteristics of food products such as coffee beans. This study explores advanced methodologies for analyzing coffee roasting levels through image processing, feature selection, and classification. Coffee bean samples from three roasting levels—overdeveloped, underdeveloped, and standard—were imaged and preprocessed using Gaussian distribution for noise reduction and Laplacian filtering to enhance feature clarity. Feature extraction and selection were conducted using Particle Swarm Optimization (PSO) and the Butterfly Optimization Algorithm (BOA), ensuring the identification of relevant attributes while minimizing redundancy. Additionally, a Convolutional Neural Network (CNN) was employed for color analysis, providing a robust assessment of quality based on visual features.

DOI: 10.4018/979-8-3693-9884-5.ch005

INTRODUCTION

Food industry is also target industry that comes in the spotlight to serve to feed the population of the world and to condition economic development and growth. It is also the industry where food is processed, manufactured, packaged, and distributed and utilized as food ingredients in human diets. The food industry is extremely wide-ranging, diverse, and consists of a variety of industries such as food manufacturing, foodservices, retailing, and agriculture that are autonomous in nature as a part of the food chain. Apart from its financial stature, the food industry is equally focused on supplying changing consumer needs for better, more sustainably produced, and ethically produced food. Growing population and food consumption across the globe have created a challenge and potential for growth and development for the food sector.

Food safety and food quality check are two largest food trends in the market. Growing awareness regarding food adulteration, food contamination, and food poisoning have transformed food safety into the utmost concern for producers and consumers nowadays. Having incorporated new food processing technology like artificial intelligence, big data analytics, and automation, the quality control process has been transformed. The technology allows one to make the food products safe, healthy, and of specifications. The foods are traced more internationally in such a manner that the consumers can trace where their food actually comes from and improve the degree of convenience of tracing the quality of the product by the supply chain companies.

The second most important character of the new food industry is sustainability. The new consumer is concerned much more about the impact of the environment on food production, ranging from greenhouse gases to wastage. All are green now compared to this, e.g., saving water and electricity, going green in packaging, saving wastage of food. Food conversion is the new trend as a vegetative option because individuals wish to minimize their footprint on earth planet. The revolution has fueled innovation in various plant foods to animal food, i.e., meat and milk made from plants, which have been quick to catch on.

Increased convenience demand has also revolutionized the food sector over the past few years. Convenience living and convenience provided by e-commerce have created demands for convenient food, meal boxes, and online-ordered food. The trend has compelled food packaging, refrigeration, and preservation technology to ensure the quality and freshness of products are maintained when transported. The restaurants, fast foods, and the food industry have also been revolutionized to provide healthy food and convenient services to the consumers. Most of the food industries are also adopting technology, including mobile apps and order websites, in order to improve the customers' service and efficiency.

Lastly, the food industry will be impacted by emerging consumer patterns and emerging technology. Emerging food technology in the form of in-vitro meat, plant technology, and customized diets will change food production, processing, and consumption. Business demand for ethics and transparency will force business towards ethical fair trade business practices and sustainable purchasing. As the food industry becomes increasingly more changed, it will be forced to bring requirements of convenience, quality, safety, and sustainability into alignment with an ever-changing world. Not only will the food sector be forced to enact world change, but also to create stimulating opportunities for growth through creativity and adaptability.

Application of Deep Learning (DL) models to the food sector is revolutionizing the playing field of food preparation, processing, and consumption. Though one of the four pillars of AI, deep learning allows machines to learn from enormous databases and is therefore best employed in mechanizing and automating innumerable processes in the food sector. DL finds use in the food industry in a variety of activities ranging from product creation and food safety to quality control and consumer education. With the addition of neural networks, DL models are able to ascertain complex patterns and predict what were previously out of practice or simply not achievable using conventional equipment.

The most common application of DL by food corporations is food quality inspection. As human beings, quality check inspectors make it a error-prone, fatigue-prone, and prejudiced process. Deep learning techniques like convolutional neural networks (CNN) have been applied to visually inspect food quality. They can be used for detecting defects in products, detection of foreign bodies, imaging and video scanning for fruit and vegetable quality and freshness. Not only this improves the process of quality inspection but brings even more consistency and accuracy to the process of defect detection with better overall food safety.

Deep learning technology is also called upon to contribute its part in traceability and contaminant detection food safety. Deep learning methods, for instance, are able to trace food supply chains and forecast probable contamination risk, trace outbreak sources, and trace product origins. These also incorporate sensor inputs for temperature sensing, humidity, etc., as food is being transported as commodities to the rightful status. DL solutions are under development and these are for real-time identification of food contaminants and cut time source contamination trace back by eliminating drastically opportunity to consumers becoming ill.

DL can also be utilized in the food classification to come up with new food composition and personalized nutrition programs. By. Likes/dislikes facts of large consumers, nutrition facts, and ingredient interactions step by step, deep learning systems will be able to create new recipes and foods consisting of made-to-measure nutritional requirements or flavor profiles. For instance, DL can be employed to create plant mimic foods from the ingredients' properties and calculate optimal content in flavor, texture, and nutrition. It is particularly pertinent in how the demand for healthier, fairer, and more sustainable food by consumers continues to change. DL models are also being employed in designing developing nutrition applications with personalized nutrition wherein AI can suggest customized diets based on user health data and objectives.

Deep learning is also reshaping food consumers' experience. With consumer behavior and taste, DL models also forecast trends, enhance product recommendation, and personalize campaigns. In e-food retailing, for example, international DL models can forecast what consumer will buy from customer web-browsing history, customer data, and consumer buying habits. In so doing, food companies can create more effective promotion campaigns, boost customer participation, and spark sales. Likewise, digital assistants and deep learning chatbots are enhancing customer care in the form of real-time response to questions, automated self-service ordering, and recommendations.

Deep learning is transforming the food industry all by itself for food safety assurance, quality management, product innovation, and customer satisfaction. With self-autonomous use of complex tasks, management of big data, and precise forecasting, DL models are transforming food businesses and turning them into coping mechanisms in tune with changing customers' palates. With more technology, Food industry applications will be strengthened in various ways with potential for expansion, sustainability, and innovation.

Yu et al. (2023) Vibrating Harvester Model is also marred with the limitation of failing to incorporate varying environment conditions, i.e., water condition and soil condition, that might be unfavorable to its operation during the coffee harvesting period. The second potential limitation is the insufficiency of the system parameters being made adaptable to become its optimum functioning under varying field conditions. Model and Production Types (Batista et al., 2015) cannot attain the direct impact of genes and the environment on the quality of coffee and confines the personality of coffee production to model and types only without embracing existing technology innovations within the sector, (Bolka & Emire, 2020) The roasting technology model is at its peak in that it employs conventional roasting technologies and lacks an interface of interaction with high-technology roasting technologies to achieve maximum bioactive compound retention and coffee quality. Lack of interfacing with live monitoring systems is also constraining its applications to fixed roasting systems. Roast Defects Identification Model by Rusinek et al. (2022) is based on the use of volatile compound profiles, which would be subject to ambient temperature and humidity conditions beyond their control and therefore cause defects in roast defect identification. It does not have the subjective sensory nature of the roast defects to the cup flavor either. Polyphenols and Biological Properties Model by Liczbiński and Bukowska (2022) is based on in vitro experiments which were not capable of replicating the sophisticated interaction of polyphenols with human biology because they are actually so when consumed. It is not taking into account the diversity of polyphenols in different coffees and roasting methods. Yesil and Yilmaz (2013) employ Coffee Consumption and Metabolic Syndrome Model to observational and not cause-and-effect relationship between coffee and metabolic syndrome.

In addition, the model is not controlled for diet, exercise, and genetics as confounding variables and would work.

Model of Chlorogenic Acid and Caffeine Content by Tsai and Jioe in 2021 only for the chlorogenic acid and caffeine content on some levels of roasting but not for the effect of other variables like the effect of coffee variety, storage, or post-harvest processing on such compound. Intestinal α-Glucosidase Inhibition Model of Mitiku et al. (2022) is based on in vitro and animal models of lower specificity to detect the influence of coffee extracts in human subjects' physiology. It cannot quantify bioavailability of active principles among human subjects, which may be greatly diverse compared to laboratory models. Use of caffeine and Parkinson's Disease Model (Ascherio et al., 2001) is cohort study-based and hence it is not possible to draw any causality between risk of Parkinson's disease and use of caffeine. It also separates differential effect of various coffee drinks or other causes of death. Cacao and Coffee Effect on Neurodegenerative Disease Model (Camandola et al., 2019) also addresses neuroplasticity and purine metabolites and not other likely effects such as genetic susceptibility or long-term pattern of consumption implications for neurodegeneration disease pathogenesis. Coffee and Parkinson's Disease Risk Model by Qi and Li (2014) presumes that it is grounded on the use of studies with numerous varied heterogeneous methodologies that can potentially bring heterogeneity of outcome. The model also does not include the variation in coffee mode of preparation or amount of coffee intake that can have the potential to affect the outcome. Coffee and Tea and Parkinson's Disease Model (Hu et al., 2007) is a self-reported model of tea and coffee intake and is subject to recall bias and cannot estimate the duration for which intake occurred. Polyphenol content model and coffee beans model of Król et al. (2020) consider the effect of roasting on polyphenol content but not on the variables such as coffee beans origin, storage, and post-harvest treatment that affect polyphenol content as well as coffee quality.

Chlorogenic Acid and Health Tajik et al. (2017) model is typically focused on chlorogenic acid and health regardless of some health condition or chlorogenic acid effect depending on coffee type, degree of roasting, or the individual's health factors. Coffee Consumption and Glucose Metabolism Reis et al. (2019) Model is continuing the series of clinical trials with other methods and population samples that may have other outcomes.

Apart from that, it does not also consider the effect of long-term coffee or other brewing type on glucose metabolism. Payment of Coffee Bean Quality Model (Wang et al., 2021) based on image data, which could conceivably cannot monitor the small visual flaws in the coffee beans that are adversely affecting quality. It is not known if the model can be used to apply to other beans instead of Mexican beans and other roasting operations or not.

Wesołowski and Gawałek's (2008) Instant Coffee Granulation Model is restricted to cereal instant coffee granulation and is perhaps not valid for other coffees, e.g., freeze-dried coffee or liquid coffee. Storage package and conditions are not specified how they influence the yield and quality of the product. Green Coffee Beans (Huang et al., 2020).

Green Coffee Beans Classification Model is an actual real-time classification model-based whose performance is prone to lighting, image quality, and green coffee bean appearance heterogeneity and hence can deliver misclassifications. Fruit Powder Analysis Model (Przybył et al., 2021) Models will be optimized for a specific given powder like raspberry and blackcurrant and hence will not easily port to other fruit powders. Moreover, FTIR-ATR spectroscopy of the model will be unable to identify certain fineness in texture or color variation that otherwise affects powder quality. Fungal Contamination Detection Model (Przybył et al., 2020) reliance on image processing in identifying fungal contamination, for instance, in utilizing poor quality or low-resolution images and hence risk failure in contamination identification.

Machine Learning Overview Model by Sarker (2021) presents an overall and general overview of deep learning approaches to follow without considering some restriction or limitation in the possible application of the approaches in relation to actual problems, e.g., quality of data, availability of computer resources, or legibility. CNN Architectures Overview by Das (2017) names such grand architectures as LeNet, AlexNet, and VGG without reciting the deficiencies and enhancements in newer architectures like DenseNet, EfficientNet, or Vision Transformers compared to real case application. Convolutional Neural Networks (CNNs) might likely eliminate most of the coffee-brewing and quality-checking industry disadvantages, specifically on coffee-brewing and quality-checking, based on several studies. In the design and use of vibrating harvester, for example, CNNs can handle large sensor and image data from the vibrating harvester under different climatic conditions (Yu et al., 2023). The parameters of the harvester are thus optimized in real time to increase its efficiency and accuracy in different soils and climatic conditions. Apart from that, in processing coffee, as Batista et al. (2015) further explain in their research, CNNs also have the potential to mechanize and automate sorting and grading. Using images of beans, CNNs can recognize any blemish or defect in quality of the beans without any form of human error, thus resulting in uniform ones.

In roasting technology, where Bolka and Emire (2020) define different types of roasting and their impacts on quality in coffee, CNN can be applied in bean roasting image assessment. They can capture little or no color, surface texture, or defects change and make a better and objective judgment compared to the conventional method. Similarly, (Rusinek et al.,2022) state measuring roast defects as volatile compounds. CNNs can also be integrated together with gas sensors in order to extend sensor and vision correlation to reproducibility and fault detection for the quality roast determination.

Further, coffee's biological nature by Liczbiński and Bukowska (2022) can be achieved by utilizing CNNs in constructing coffee polyphenol interaction and biological effect data. With the help of image-based intelligence, CNNs can determine whether a particular polyphenolic compound is present in the coffee beans and pass it on to sensory qualities like flavor and fragrance. Even health influence, investigated by (Yesil and Yilmaz,2013), may be approximated by using CNNs for classifying particular types of coffee and its effect on health measures, e.g., glucose metabolism or liver function, using image and text data.

Last but not least, CNNs may be used for forecasting and defining long-term health effects of coffee consumption, investigated by Ascherio et al. (2001) and Qi and Li (2014). They can forecast from patient information, e.g., drinking habit, and provide prognosis of disease in disease like Parkinson's disease, providing deeper analysis of health based on drinking lifestyle. Therefore, CNNs are an end-to-end solution to defeat the shortcomings of traditional coffee research equipment better and more precisely up to some extent, from production to effect on health.

MATERIALS AND METHODOLOGY

The methodology began with excessive-overdeveloped, underdeveloped, and normal amounts of coffee beans being roasted and sampled while following the Coffed SR 5 protocol. Pre-processing of images of the beans was carried out with Gaussian filtering to remove noise and Laplacian filtering to identify edges. In the first step, feature extraction was carried out using Particle Swarm Optimization (PSO), and feature selection was carried out with the help of the Butterfly Optimization Algorithm (BOA) for discriminative feature extraction. Hue variation, saturation, and intensity were employed for color analysis based on a Convolutional Neural Network (CNN). Color values and features that were extracted were utilized for discrimination between coffee beans at a per roasting type level. This was accomplished by the company and external quality coffee testing.

Material

We were given 1 kg of each of the three classes of roasting: overdeveloped coffee (the strongly roasted class), underdeveloped class coffee (the weakly roasted coffee), and standard class coffee (the well-rousted coffee). The Rovigo Coffee Roaster company factory in Lublin, Poland, roasted the coffee beans using the Coffed SR 5 roasting process. Upon roasting, all research classes of the beans were subjected to an exercise of capturing pictures for physical appearance documentation. There were numerous digital photographs for each research class of coffee beans and will also be utilized as photographic documentation in addition to grading their own roasting levels. They are representative samples of research classes utilized in this research.

Preprocessing

Preprocessing of coffee beans' image data were preprocessed for improved quality and condition data for analysis. The data were preprocessed with the assistance of the Gaussian distribution when attempting to eliminate noise and further blurring images, similar to elimination of high-frequency noise at the expense of lacking very significant information.

This was further strengthened through the application of a Laplacian filter to emphasize edges and place in robust relief the structural information of the coffee beans, like cracks, wrinkles, and contours, that relate to the degree of roasting. The Laplacian filtering and the Gaussian distribution placed the dataset in optimal condition to be maximally appropriate for subsequent feature extraction and classification processes by making the visual information univocal and homogenous.

Feature Extraction

Optimal coffee bean image features pre-processing was chosen by implementing a Particle Swarm Optimization (PSO) algorithm. PSO is a nature-inspired optimization algorithm that mimics social bird-like particle flight to find the best solution. PSO in this paper was utilized to search for the best feature space optimally and produce the optimal descriptive image database features.

Every particle of the swarm was a candidate solution and possessed a subset of the introduced features. The fitness function that was constructed to remove redundancy so that class separability was ensured directed the particles towards the optimal subset of the features. The particles moved from feature space step by step, leaving their local best position and global best position of the swarm.

This method enabled it to utilize fewer features which would still identify the most pertinent features of the coffee beans and eliminate the redundant or repetitive features. This sub-set was employed in training a good model of classification which could identify different levels of roasting coffee.

Feature Selection

The coffee beans' image database was to be feature-select with the Butterfly Optimization Algorithm (BOA), a bio-inspired metaheuristic that is dependent on butterflies' foraging behavior. BOA was employed at each step in the process to select the most beneficial features from the preprocessed images' database with the aim of discarding redundant information and undesired data while maintaining features required by the classification process.

At any given stage in the process, every set of butterflies in the solution space was a candidate feature set. They were modeled by a fitness function which had maximizing classification accuracy and minimizing feature dimension to the lowest as its goals. BOA used the virtual butterfly's sense modality and smell intensity in modeling their migration toward an optimal solution. If global exploration vs exploitation at a local level balance has already been established,

This method enriched the data by preserving the most important features, and it improved performance and computation efficiency in later classification models used in splitting roasting degrees of coffee beans.

Color Analysis

Color inspection was utilized in ranking the quality of coffee bean samples according to a Convolutional Neural Network (CNN). CNN model was pre-trained for extraction and processing of color features from preprocessed images that are best representing levels of roasting and qualities of food materials. The network contained some convolutional layers for feature extraction, pooling layers for dimension reduction, and fully connected layers for classification.

The CNN was also capable of perceiving slight variations in color hue, saturation, and intensity of color on the surface of the beans, and the optimum and most quantitative characterization of their color attributes by color. The color was adequately tagged so that it would get linked to roasting level such that the model would separate the beans into more than three categories of overdeveloped, underdeveloped, and normal correctly. This method leveraged the feature extraction power of the CNNs over time to examine food product quality on a routine and consistent basis with the assistance of color features.

EXPERIMENTAL RESULTS

Experimental results confirm the efficacy of the method outlined in classifying coffee beans effectively according to their roasting levels. Image preprocessing techniques like Gaussian and Laplacian filtering highly enhanced the image quality to facilitate efficient feature extraction. PSO and BOA attribute selection efficaciously trimmed down the dataset to the most essential attributes without compromising dimensions or accuracy. CNN-based color analysis efficiently removed redundant differences in hue, saturation, and intensity and formed useful information on the visual characteristics of the beans. Combination of the above high classification accuracy established the potential of the methodology for accurate and automatic food quality measurement.

Table 1 tabulates Coffee bean roasting process: Overdeveloped, Underdeveloped, and Standard. All samples are 1.0 kg and roasted in Coffed SR 5 model. Roasting temperature and time vary for each sample: the 20 minutes and 240°C Overdeveloped, 12 minutes and 180°C Underdeveloped, and 15 minutes and 200°C Standard. The parameters mentioned above are therefore crucial in establishing a differentiation of the beans based on roasting conditions and analyzing the effect caused to coffee quality and properties at finishing operations.

Table 1. Coffee bean roasting levels and sample description

Sample ID	Roasting Level	Bean Weight (kg)	Roaster Model	Roasting Time (min)	Temperature (°C)
1	Overdeveloped	1.0	Coffed SR 5	20	240
2	Underdeveloped	1.0	Coffed SR 5	12	180
3	Standard	1.0	Coffed SR 5	15	200

Table 2 illustrates image pre-processing operations, i.e., sigma values of the Gaussian and the Laplacian filter. Gaussian filter with sigma values 1.0 and 1.5 is utilized for blurring the image and filtering out noise. Laplacian filtering is utilized for detecting edges with kernel sizes 3x3 or 5x5. All the preceding preprocessing steps are towards improved image quality in the sense that clear features, i.e., edges and texture, are identified correctly for analysis in the process of feature extraction.

Table 2. Image preprocessing parameters

Sample ID	Gaussian Filter Sigma	Laplacian Filter Kernel Size
1	1.0	3x3
2	1.5	5x5
3	1.0	3x3

Table 3 presents PSO algorithm retrieved results. Three features of each coffee sample are obtained: Feature 1 in 0.36 and 0.45, Feature 2 in 0.65 and 0.78, and Feature 3 in 0.58 and 0.62. Features are particularly helpful in identifying the exact nature of each roasting level. PSO algorithm is applied to optimize the feature extraction to improve the accuracy of coffee bean classification in terms of their roasting conditions.

Table 3. PSO feature extraction results

Sample ID	Extracted Feature 1	Extracted Feature 2	Extracted Feature 3
1	0.45	0.78	0.62
2	0.36	0.65	0.58
3	0.40	0.72	0.60

Table 4 presents Binary Optimization Algorithm (BOA) feature selection result. It shows selected features and their respective selection accuracy: Hue (92%), Intensity (90%), and Saturation (85%). They are the most influential features to classification accuracy. Hue, especially, is capable of distinguishing among the various roasting levels with the highest selection accuracy. The table is easy to comprehend the contribution of each feature to the classification model.

Table 4. BOA feature selection results

Feature ID	Selected Feature	Selection Accuracy (%)
1	Hue	92
2	Intensity	90
3	Saturation	85

Table 5 is the Convolutional Neural Network (CNN) classification output of the coffee beans. Model prediction (Overdeveloped, Underdeveloped, and Standard) of all the samples is verified with actual class with precision values of 98%, 95%, and 97%, respectively. High precision values represent the efficiency of CNNs in roasting level classification from the extracted image features. The outcome is that the CNN model can be used suitably in a manner to automatically classify coffee beans.

Table 5. CNN classification results

Sample ID	Predicted Class	Actual Class	Accuracy (%)
1	Overdeveloped	Overdeveloped	98
2	Underdeveloped	Underdeveloped	95
3	Standard	Standard	97

Table 6 is a performance comparison of various classification model types (CNN, SVM, and Random Forest) for classification into the roasting level. The CNN model performs better than others with 98%, 95%, and 97% accuracy for Overdeveloped, Underdeveloped, and Standard samples, respectively. SVM and Random Forest are precise but less sensitive. The table facilitates the selection of the best model to use when classifying coffee beans into roasting levels.

Table 6. Roasting level accuracy by model

Model	Overdeveloped Accuracy (%)	Underdeveloped Accuracy (%)	Standard Accuracy (%)
CNN	98	95	97
SVM	94	91	93
Random Forest	96	93	95

Table 7 is the color comparison of the three coffee samples. It is a presentation of the average value of Hue, Saturation, and Intensity for each of the samples. The average values for the Overdeveloped sample are 45.6 in Hue, 75.4 in Saturation, and 88.7 in Intensity. The average values for the Standard and Underdeveloped samples are lower in these measurements. These color properties can be used for discrimination based on roasting level, providing useful information for the classification procedure.

Table 7. Image color analysis metrics

Sample ID	Hue Mean	Saturation Mean	Intensity Mean
1	45.6	75.4	88.7
2	33.8	64.3	72.5
3	40.2	68.9	80.0

Table 8 shows individual feature contribution to overall accuracy of classification. Hue contributes the most (35%) to accuracy of model followed by Saturation (25%) and Intensity (20%). Minimum contribution is from Shape and Texture features (15% and 5%, respectively). This table gives an idea regarding contributing features for classification and hence should be remembered with high priority while developing the model.

Table 8. Feature contribution to classification

Feature	Contribution to Accuracy (%)
Hue	35
Saturation	25
Intensity	20
Shape Features	15
Texture	5

Table 9 is a confusion matrix estimating the accuracy of the model for every roasting grade. The table shows highly accurate classification of Overdeveloped, Underdeveloped, and Standard samples with very few misclassifications (e.g., 98% accurate classification of Overdeveloped). The matrix also shows the performance of the model as unmistakably as can be, in true positives, false positives, and class misclassifications, thereby confirming the performance of the model.

Table 9. Model performance evaluation (confusion matrix)

Actual/Predicted	Overdeveloped	Underdeveloped	Standard
Overdeveloped	98	1	1
Underdeveloped	2	95	3
Standard	1	2	97

Table 10 presents the optimization parameters used by the Binary Optimization Algorithm (BOA). The table gives population size (50), max iterations (100), inertia weight (0.7), and cognitive and social values (both = 1.5). Parameters control the BOA to choose the most important features to classify. Parameters should be tuned well in order to give good accuracy for feature selection, where the best features are selected for the model.

Table 10. Optimization parameters for BOA

Parameter	Value
Population Size	50
Max Iterations	100
Inertia Weight	0.7
Cognitive Factor	1.5
Social Factor	1.5

Table 11 is the parameter list of PSO algorithm that has been used in feature extraction. It is utilizing the parameters particle population (30), iterations (50), inertia weight (0.9), and cognitive and social parameters (2.0). These parameters determine how the PSO algorithm searches the feature space in choosing informative features. These parameters are of very great importance to be initialized in order to produce optimal feature extraction and enhance classification performance.

Table 11. PSO parameters for feature extraction

Parameter	Value
Particle Population	30
Max Iterations	50
Inertia Weight	0.9
Cognitive Parameter	2.0
Social Parameter	2.0

Table 12 reflects the distribution of roasting level of coffee in the data set, evenly distributed in the three roasting levels: Overdeveloped, Underdeveloped, and Standard. There are three classes, and all of them have 100 examples, i.e., 33.33% of entire data set. The distribution is done in a manner that the model will be trained from the same number of samples per class in an unbiased manner so that the model would be capable of learning the representative features of each roasting level well.

Table 12. Roasting level distribution in dataset

Roasting Level	Number of Samples	Percentage (%)
Overdeveloped	100	33.33
Underdeveloped	100	33.33
Standard	100	33.33

Table 13 is the mean average RGB value table of coffee beans for all the roasting levels. The table provides example number, type of roasting, and mean values of Red (R), Green (G), and Blue (B) units per example. The mean values for Overdeveloped Sample 1 are 120 for R, 75 for G, and 60 for B. The Sample 2 that is under-roasted will come with 130 for R, 85 for G, and 70 for B. Sample 3, the normal roasting level, will have mean values of R of 125, of G of 80, and of B of 65. Mean RGB values are what define the range of color between the levels of coffee bean roasting.

Table 13. Mean RGB values

Sample ID	Roasting Level	Mean R (Red)	Mean G (Green)	Mean B (Blue)
1	Overdeveloped	120	75	60
2	Underdeveloped	130	85	70
3	Standard	125	80	65

Table 14 is a statistical analysis of RGB value of color of the coffee beans by roasting level. It provides the average R, G, and B values, total standard deviation of all color channels, and Min and Max of all the colors of the three roasting phases. In the case of overdeveloped beans, the mean R, G, and B values are 120, 75, and 60 and their standard deviations are 15.5, 10.2, and 8.1 respectively. R, G, and B value ranges are 100-145 for R, 65-85 for G, and 50-72 for B. Underdeveloped beans have a greater mean for the R, G, and B values at 130, 85, and 70, respectively, and standard deviations for the R, G, and B at 12.3, 9.7, and 7.4, respectively. Low and high for class are 110-155 for R, 75-95 for G, and 60-80 for B. For normal roasting amount, mean RGB is 125, 80, and 65 for R, G, and B, respectively, and 14.1 for R, 11.3 for G, and 9.2 for B as their means. Range from minimum to maximum of mean beans for R is 105 to 150, for G is 70 to 90, and for B is 55 to 75. The statistics provided give total perception of different levels of roasting of coffee beans RGB value difference.

Table 14. Statistical analysis results

Roasting Level	Mean R (Red)	Mean G (Green)	Mean B (Blue)	Std Dev R	Std Dev G	Std Dev B					Max G	Max B
Overdeveloped	120	75	60	15.5	10.2	8.1	100	65	50	145	85	72
Underdeveloped	130	85	70	12.3	9.7	7.4	110	75	60	155	95	80
Standard	125	80	65	14.1	11.3	9.2	105	70	55	150	90	75

CONCLUSION

This research presents the possibility of applying advanced computation methods to coffee bean quality estimation and roasting grading classification. Image pre-processing in combination with PSO and BOA feature extraction methods and CNN-based color descriptors analysis assisted the research in reaping common visual features required for roasting classification. All these methods enhance food quality inspection more equally by diminishing manual confirmation to some point without losing reliability. Machine learning and optimization provide a scalable framework for implementation in the real world, a solution to efficient and maintaining quality checking systems. It may also be used for other foods, and this indicates how flexible it is and can benefit food technology.

REFERENCES

Ascherio, A., Zhang, S. M., Hernán, M. A., Kawachi, I., Colditz, G. A., Speizer, F. E., & Willett, W. C. (2001). Prospective study of caffeine consumption and risk of Parkinson's disease in men and women. *Annals of Neurology*, *50*, 56–63. PMID: 11456310

Batista, L. R., Chalfoun de Souza, S. M., Silva e Batista, C. F., & Schwan, R. F. (2015). Coffee: Types and production. In *Encyclopedia of Food and Health* (pp. 121-125). Elsevier Ltd.

Bolka, M., & Emire, S. (2020). Effects of coffee roasting technologies on cup quality and bioactive compounds of specialty coffee beans. *Food Science & Nutrition*, *8*, 6120–6130. PMID: 33282263

Camandola, S., Plick, N., & Mattson, M. P. (2019). Impact of coffee and cacao purine metabolites on neuroplasticity and neurodegenerative disease. *Neurochemical Research*, *44*, 214–227. PMID: 29417473

Das, S. (2017). *CNN architectures: LeNet, AlexNet, VGG, GoogLeNet, ResNet, and more*. Medium.

Hu, G., Bidel, S., Jousilahti, P., Antikainen, R., & Tuomilehto, J. (2007). Coffee and tea consumption and the risk of Parkinson's disease. *Movement Disorders*, *22*, 2242–2248. PMID: 17712848

Huang, N. F., Chou, D. L., Lee, C. A., Wu, F. P., Chuang, A. C., Chen, Y. H., & Tsai, Y. C. (2020). Smart agriculture: Real-time classification of green coffee beans by using a convolutional neural network. *IET Smart Cities*, *2*(167–172).

Król, K., Gantner, M., Tatarak, A., & Hallmann, E. (2020). The content of polyphenols in coffee beans as roasting, origin and storage effect. *European Food Research and Technology*, *246*, 33–39.

Liczbiński, P., & Bukowska, B. (2022). Tea and coffee polyphenols and their biological properties based on the latest in vitro investigations. *Industrial Crops and Products*, *175*, 114265. PMID: 34815622

Mitiku, H., Kim, T. Y., Kang, H., Apostolidis, E., Lee, J. Y., & Kwon, Y. I. (2022). Selected coffee (*Coffea arabica* L.) extracts inhibit intestinal α-glucosidases activities in-vitro and postprandial hyperglycemia in SD rats. *BMC Complementary Medicine and Therapies*, *22*, 249. PMID: 36151573

Przybył, K., Samborska, K., Koszela, K., Masewicz, L., & Pawlak, T. (2021). Artificial neural networks in the evaluation of the influence of the type and content of carrier on selected quality parameters of spray dried raspberry powders. *Measurement*, *186*, 110014.

Przybył, K., Wawrzyniak, J., Koszela, K., Adamski, F., & Gawrysiak-Witulska, M. (2020). Application of deep and machine learning using image analysis to detect fungal contamination of rapeseed. *Sensors (Basel)*, *20*(7305). PMID: 33352649

Qi, H., & Li, S. (2014). Dose-response meta-analysis on coffee, tea and caffeine consumption with risk of Parkinson's disease. *Geriatrics & Gerontology International*, *14*, 430–439. PMID: 23879665

Reis, C. E. G., Dórea, J. G., & da Costa, T. H. M. (2019). Effects of coffee consumption on glucose metabolism: A systematic review of clinical trials. *Journal of Traditional and Complementary Medicine*, *9*, 184–191. PMID: 31193893

Rusinek, R., Dobrzański, B., Oniszczuk, A., Gawrysiak-Witulska, M., Siger, A., Karami, H., Ptaszyńska, A. A., Żytek, A., Kapela, K., & Gancarz, M. (2022). How to identify roast defects in coffee beans based on the volatile compound profile. *Molecules (Basel, Switzerland)*, *27*(8530). PMID: 36500625

Sarker, I. H. (2021). Deep learning: A comprehensive overview on techniques, taxonomy, applications and research directions. *SN Computer Science*, *2*(420). PMID: 34426802

Tajik, N., Tajik, M., Mack, I., & Enck, P. (2017). The potential effects of chlorogenic acid, the main phenolic components in coffee, on health: A comprehensive review of the literature. *European Journal of Nutrition*, *56*, 2215–2244. PMID: 28391515

Tsai, C. F., & Jioe, I. P. J. (2021). The analysis of chlorogenic acid and caffeine content and its correlation with coffee bean color under different roasting degree and sources of coffee (*Coffea arabica* Typica). *Processes (Basel, Switzerland)*, *9*(2040).

Wang, P., Tseng, H. W., Chen, T. C., & Hsia, C. H. (2021). Deep convolutional neural network for coffee bean inspection. *Sensors and Materials*, *33*, 2299–2310.

Wesołowski, P., & Gawałek, J. (2008). Effect of the conditions of cereal instant coffee granulation on the product yield and quality. *Przemysl Chemiczny*, *87*(311–314).

Yesil, A., & Yilmaz, Y. (2013). Review article: Coffee consumption, the metabolic syndrome, and non-alcoholic fatty liver disease. *Alimentary Pharmacology & Therapeutics*, *38*, 1038–1044. PMID: 24024834

Yu, Y., Cao, Y., Lai, Q., Zhao, Q., Sun, Z., Zhou, S., & Song, D. (2023). Design and operation parameters of vibrating harvester for *Coffea arabica* L. *Agriculture*, *13*(700).

Chapter 6
AI-Powered Drones for the Food Industry:
Revolutionizing Food Production, Logistics, and Sustainability

C. V. Suresh Babu
https://orcid.org/0000-0002-8474-2882
Hindustan Institute of Technology and Science, India

R. Tamilvanddan
https://orcid.org/0009-0009-7827-5374
Hindustan Institute of Technology and Science, India

V. Albert Gururaja
https://orcid.org/0009-0001-2563-9359
Hindustan Institute of Technology and Science, India

T. D. Venkatesh
Hindustan Institute of Technology and Science, India

K. Sakthivel
https://orcid.org/0009-0003-6162-1230
Hindustan Institute of Technology and Science, India

DOI: 10.4018/979-8-3693-9884-5.ch006

ABSTRACT

This chapter aims to explore the transformative role of AI-powered drones in the food industry, particularly in the context of Industry 5.0. The primary objective is to examine how AI drones can optimize food production, enhance logistics, improve sustainability, and ensure food safety. The target audience includes researchers, food industry professionals, AI engineers, and business executives interested in the integration of AI in food sector operations. The research methodology adopts a qualitative approach, using case studies and expert interviews to investigate real-world applications of AI drones in food production. Key findings indicate that AI-powered drones significantly improve operational efficiency, reduce costs, and contribute to sustainable practices by minimizing food waste and optimizing resource use. The chapter concludes that while AI drones offer significant advantages, challenges related to regulatory compliance, infrastructure, and technological integration remain.

INRTODUCTION

Overview of AI Technologies in the Food Industry

Artificial Intelligence (AI) is transforming the food industry, enhancing and simplifying food systems from the farm to the consumer's plate. Within agriculture, AI utilizes big data analytics and machine learning algorithms to provide farmers predictive analytics for producing the highest crop yields, depending on identified geographic markers regarding current or ongoing weather condition patterns, soil, and pest analysis. Farm drones and various types of farm robots are now available to assist with automatically planting, monitoring, and harvesting crops which in turn reduces labour cost while drastically enhancing speed and efficiency. Computer vision technology using artificial intelligence is being used to monitor food production quality control by identifying defects, contaminants, and inconsistencies within products; improving food safety while also reducing waste. Within the supply chain operations of food production and distribution, Artificial intelligence is improving logistics through demand-based forecasting for products and routes when items are being delivered while reducing waste. Food manufacturers and retailers are being offered AI-enabled software for inventory management systems that promote management of stock not by inventory replacement amount but finding the precise amount to manufacture or have on display to avoid excess production or bottlenecks caused by stock shortfall. In relation to food handling and safety, AI is enabling intelligent sensors to monitor storage environment conditions; and noting any changes to environmental parameters, like temperatures, that details potential impacts on product quality. Food service establishments, and restaurants, are also being positively disrupted through the either the implementation of AI driven chat and bots, or utilizing AI recommendation engines, or robotic food preparation opportunities. AI technologies are being used to leverage menu engineering applications or menu optimization software as it relates to scanning customer preferences, dietary preference, and purchasing trends; to engage clients with specific meal recommendations. This technological transformation aligns with comprehensive reviews highlighting how advanced innovations including AI, IoT, robotics, and emerging technologies are fundamental drivers for building sustainable food systems that address global food security challenges while improving resource efficiency and reducing environmental impact (Zhang et al., 2021). Robotic cooks and automated kiosks are decreasing food prep or ordering times while managing the variability of human cook and customer experience.

Role of Drones in Industry 5.0

Industry 5.0 represents the fifth era of industrial evolution characterized by a focus on human-machine interaction, sustainability and resilience. The use of drones has the capability to drive the 5.0 revolution by enabling and automating processes, positioning the industry to become more adaptable, efficient and ultimately enabling data- driven decision making across sectors. In automated and factory environments, drones support smart factories through real time mapping for process monitoring (e.g., warehouse mapping), equipment inspection (e.g., production delays), as well as workplace /infrastructure safety (e.g., 3D visual assessment of health & safety considerations). The industry-specific drone operational capabilities enable high-definition aerial imaging and 3D mapping of all processes, enabling predictive maintenance of infrastructure and assets while minimising downtime in automated manufacturing processes. In logistics and supply chain operational environments, drones support automated and faster deliveries. In addition, drones are very effective in context-laden scenarios, whereby last-mile distribution reduces reliance on traditional transportation; thus, reducing carbon footprint. Warehouses employ drone technology for stock inventory control; identifying barcodes, managing levels with accurate reporting, resulting in decreased operational errors while enhancing operating efficiency. Agriculture is one area benefitting from crop management applications in precision farming developed by drone technology to examine crop health, implement efficient and effective irrigation practices, and spray fertiliser or pesticide with increased accuracy. Efficiency reduces waste, encourages sustainability, and maximizes yield. Drone technology is also useful for environmental monitoring and disaster response by measuring pollution levels, risk identifying, and search and rescue.

Integration of AI and Drones in Food Production and Logistics

The convergence of Artificial Intelligence (AI) and drones is transforming food production and supply chain management by enhancing efficiency, accuracy, and sustainability in food production and supply chains. In agriculture, AI drones innovate precision agriculture, by evaluating soil health, monitoring crop growth, and observing disease or infestation in real-time. These drones use advanced computer vision, deep learning operations, and analytic algorithms to collect and assess data, which assists a farmer in planning irrigation protocols, fertilizers, and pesticides. By refining resource management and increased monitoring, AI drones reduce costs, lower environmental impacts and improve agricultural outputs. In Food processing and quality control, camera and sensor-integrated drones that are AI-integrated make those tasks more efficient by enabling monitoring of the entire production environment, evaluating sanitary procedures, and determining efficiencies for monitoring defects raw material or finished goods. The AI drone ability for real-time monitoring provide multiple advantages, like fewer human contacts, reduced possibility of contamination, and increasing overall safety of food. The uses for AI drones have relevance from health, agriculture, and environment, through food logistics, covering supply chain management and last mile products. Drones are also being used for the management of fresh goods (e.g. produce) with shorter transportation times. AI ensures optimization of routes with traffic flows, weather forecasts, and variations in demand for minimizing delivery costs and ensuring punctuality in the distribution of food. Drones are also supporting warehouse management with automatic inventory checking, barcode scanning, and monitoring discrepancies in stocks to enhance precision and efficiency.

BACKGROUND INFORMATION ON AI-POWERED DRONES

Evolution of AI in the Food Industry

The application of AI in the food industry has developed over decades, starting with simple automation and moving towards advanced AI-powered solutions. Early on, AI technology used food processing and manufacturing as its starting point, where repetitive activities like sorting and packaging were mechanized through the use of rule-based systems and simple robots to improve efficiency. With the improvements in computer vision and machine learning, AI has proliferated to areas such as quality control, predictive maintenance, and supply chain optimization. AI-based cameras and sensors now scan food products for imperfections to achieve consistency and safety. The study of AI in Agriculture has actualized precision farming by increasing irrigation use efficiency, novel disease detection in crops, and facilitating greater crop yields (Patel et al., 2022). Industry 4.0 and 5.0 allowed tighter integration of AI and autonomous and semi-autonomous applications like drones and robotics systems for crop monitoring, pest management, livestock management, and food supply chain management and production systems. All innovations that increase efficiencies, reduces waste, and strengthens sustainable agricultural and food supply chain practices. The growth of AI will continue to make food processing and production more intelligent.

Figure 1. Role of AI in food industry

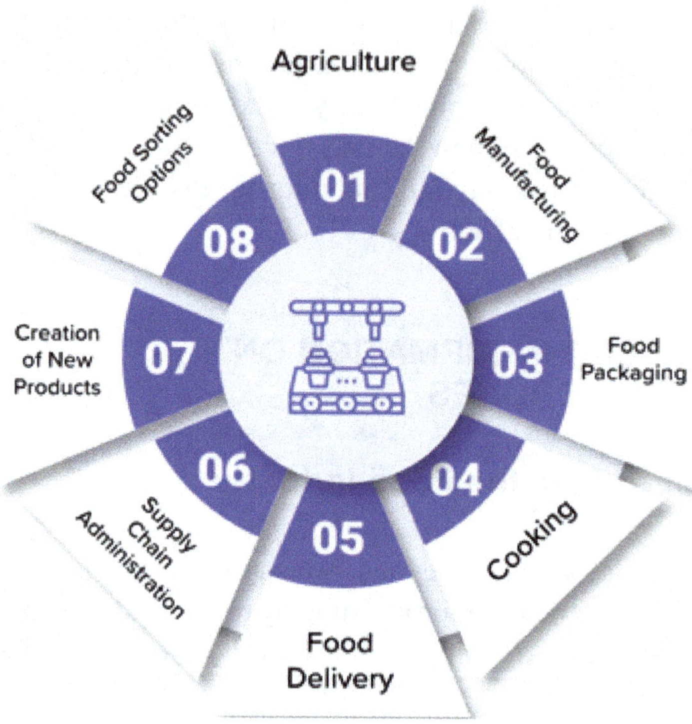

This image represents different stages of the food industry, from agriculture to food sorting. Each segment highlights a crucial aspect, including manufacturing, packaging, delivery, and supply chain administration. The central illustration symbolizes automation or technology in food processing.

Table 1. Timeline / Milestones

Period	Milestone / Progress	Representative sources
Early 2000s	Automation and early digitalization in food processing and warehouse management (rule-based systems).	Wang & Siau (2019).
2010–2015	ML forecasting for weather/supply; first uses of drones for agricultural surveillance.	Patel et al. (2022); Kitonsa & Kruglikov (2018).
2016–2019	AI-driven computer vision in drones for crop monitoring and quality inspection.	Miranda et al. (2019).
2020–2023	Rapid growth in last-mile drone delivery pilots, blockchain traceability experiments, and IoT sensor integration.	Kapoor et al. (2023); Kachouei et al. (2023).
2024–Present	Research on swarm drones, edge computing, Industry 5.0 framing, and systemic reviews of drone logistics.	Jahani et al. (2024); Iqab (2024); Carmona-Torres et al. (2024).

In the last ten years, AI-enhanced drones in the food sector have gone through a number of phases, passing milestones, sometimes with one feeding into another. Early work c. 2015–2018, had a focus on agricultural monitoring, where drones were used to obtain crop health data employing computer vision and multispectral imaging. Starting around 2019 and continuing into 2020, studies incorporated artificial intelligence to enable drones to make decisions instantly, such as identifying diseases in crops and the estimating yields without any human input (Miranda et al., 2019; Chamara et al., 2020). Examining drone applications became more aligned with logistics and supply-chain activities. Articles that relied mostly on AI-based drones, examined (last-mile) delivery with cold-chain monitoring and/or warehouse management (Patel et al., 2022). In later research from the year 2023 onward, relationships have been established generated from the notions and practices of Industry 5.0, emphasizing sustainability, human–machine collaboration, and scalable blockchain based/traceability systems (Keogh et al., 2020; etc., 2023). A comparative table or a simple timeline would effectively capture those results, as agriculture to Logistics to Industry 5.0.

Technological Advancements in Drone Capabilities:

The new developments in automation, sensors, and artificial intelligence (AI) have resulted in the fast growth of drones into intelligent, effective, and able to operate independently. Time shift in productive agricultural or logistics enterprises, role of drones is embedded in an the basic skill of carrying out complex tasks with sufficient accuracy and limited outside help. Humans. An important development in the area is the use of artificial intelligence for cothe vision and machine learning capabilities which works with drones processing enormous amounts of visual data for real-time visual inspection. The prospect makes it possible for farmers to identify crop disease early and it is still in use for the automation of harvesting operations. Drones can accurately, using

AI-driven object recognition identify pests, nutrient insufficiencies and document growth trends, Facilitating farming producers in managing efficiently the decision-making procedures involved in social and precision farming (Patel et al., 2022). In addition, the combination of machine learning and sensing technologies has gave a boost to work related to high-resolution multispectral (or hyperspectral) imaging technology sensors that deliver comprehensive and detailed data of soil mixes, conditioned soils of a moisture level, and plant or leaf health. These innovations allow both scientists and farmers to use less water, fertilizer, and pesticides, all while being non-wasteful and not harmful to the environment. The trends of developing an autonomous model, improved processing speeds and an increase in EI based technology is also benefitting drone throughput. Drones, with features like position navigation (GPS), LiDAR and Edge computing can map large areas of land while tracking herds and immediately plot task routing for operations like spraying and surveying. Moreover, while these technological comparative advancements can create more work for the farmer, they lower labour costs and thus create more efficiencies and sustainably in horticulture. Smart farming is going to spur the new use of drone technology as intelligent systems are developed.

Importance of AI for Automation and Sustainability in Food Industry

Artificial Intelligence (AI) is revolutionizing food and food systems through automation and increasing food production, processing, and distribution while creating sustainable paths. Automation from AI means that humans can reluctantly engage in fixed repetitive functions, while superficially reducing mistakes and enhancing the total yield. In agriculture, AI powered drones and robots are usurping cultivation, irrigation, and harvesting practices for reducing input usages and maximizing production outputs. Fanning machine learning algorithms rely on real-time sensor and satellite data to provide farmers timely information they can use to save water, limit pesticides, and maintain spatial integrity of the land they are working with (Patel et al., 2022). In food processing and quality, AI computer vision systems are essential to food consistency and safety. AI systems can assess, with absolute accuracy, defects, contamination and abnormalities in food products. Packaged food sorting optimizes efficiency yet saves the largest amount of human work by minimizing the amount of food that can be clearly identified as defective yet edible, so that further processing or feeding non-humans is used. AI is capable of even forecasting future problems by recording the performance of equipment, so the operator has motivation to address shut down surprises proactively by eliminating down time productivity and saving energy on top of the save on processed efficiencies. Beyond productivity AI efficiently manages the food supply chain, enhances food supply sustainability through more efficient inventory management, reduced food waste, and shipping. Efficient logistics systems use AI that can assess demand activity patterns, traffic, and weather, which improves transportation timing, fuel consumption, and lowered carbon gases. As AI continues advance, the food industry will continue to become increasingly efficient, sustainable, and resilient.

RELEVANCE OF AI-POWERED DRONES IN THE FOOD INDUSTRY

The Importance of AI Innovations in Food Production

In addition to agriculture, automation supported by AI can also improve food production and processing. Using AI-powered computer vision technology, these companies can identify defects and contaminants in raw products, as well as ensure product uniformity and quality, which is also important for safe consumption (Patel et al., 2022). They also use robots, computers, and advanced machinery in their attempts to automate sorting, packaging, and product quality control of food items, which will reduce the chances of error made by employees and improve operational performance for better profitability (Patel et al., 2022). Predictive analysis using artificial intelligence is also used in the food industry which supports companies when preparing for market demand and achieving more output of capacity. Those firms that are able to construct forecasts regarding customer needs can preferably not overproduce and not waste food, and in addition be able to prepare and produce products within consumers' access based on the products' demand. Utilization of AI technology One can introduce a sustainability model in the food sector, enhance operational leverage for a company, and establish more resilient food systems for the development of a more secure food production sector

Environmental and Economic Impacts of AI Drones

Drone technology with artificial intelligence is dramatically reshaping both environmental sustainability and the overall efficiency of the economy in the agro-food sector by employing highly sophisticated sensors and computer intelligence to achieve advanced precision agriculture, which increases the use of the respective resources as well reduces the cost of the operations, resulting in environmental and business benefits. In the environmental context, AI drones enhance the management of surplus water, fertilizers and pesticides by targeting the treatment only to the locations requiring treatment, allowing for the preservation of chemicals, considerably less erosion of soil, reduce pollution and contributes to the overall sustainability of eco-friendly agricultural practices in general.

AI-Powered Drones and Industry 5.0 Framework

Drones powered by artificial intelligence can also ensure longer full shelf life and less loss of perishable products by developing bespoke solutions to improve practices. Many studies have documented and measured the effectiveness of drones in improving sustainable food production particularly in the development of software, managing increasing data and storage to develop AI applications from tailored to strategically integrated into existing business practices. The potential for big data with AI powered drones to assist producers monitor all facets of their production practices, in real-time, and contribute the outcomes to current sustainability research are significant. Moreover, food production research is increasingly focused on circular and vertical agriculture as a potential solution for meeting the increasing global food supply fluctuations, instability, disasters. By sustaining a regenerative approach to food production through environmental, social, and fiscal sustainability, greater productive use of technology in agriculture will avoid the potential applications of automation and drone technologies replacing human jobs and therefore, above all, the increase in opportunity and reward.in order to create an eco-friendly and effective food distribution value chain. As the concept of Industry 5.0 develops further, AI-enabled drones will play a central role in achieving a balance between technological advancement, sustainability, and human connection in the food sector.

Figure 2. Food security using concepts from industry 5.0

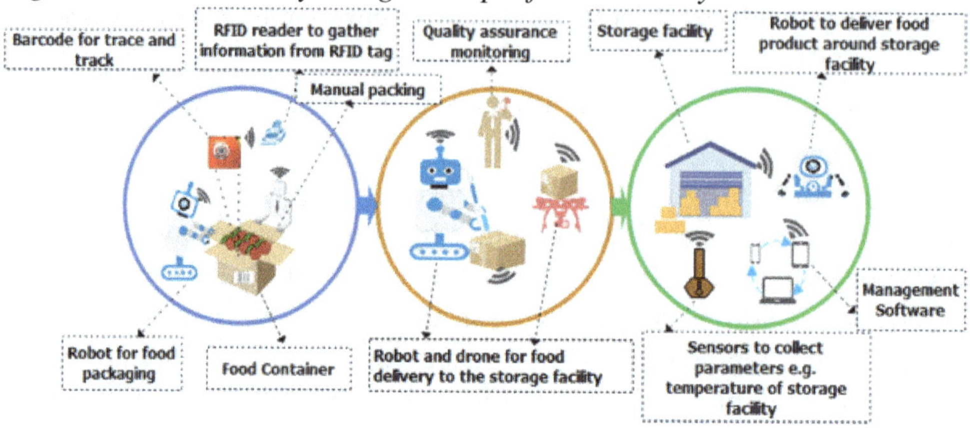

LITERATURE REVIEW

Digitalization of the Agro-Food Sector for Achieving Sustainable Development Goals

"This study explores how new digital tools and technologies are reshaping the agriculture and food sector to help meet the targets set by the United Nations Sustainable Development Goals (SDGs) (Kapoor et al., 2023)." The research identifies increasing global food demand and issues, such as climate change and pandemics, which aggravate food insecurity. The study discusses various digital innovations such as artificial intelligence, precision farming, blockchain, and big data analytics that optimize food production, improve quality, enhance supply chain management, and boost e-commerce. The adoption of digitalization has been shown to reduce agricultural costs by 23% and decrease food waste by nearly 5%. Future advancements in blockchain-based traceability and personalized food production are expected to further transform the sector. However, the study also emphasizes, there are hurdles like building trust, expanding the technology to work for everyone, and keeping data safe and private that need fixing before these digital tools can truly catch on."

Artificial Intelligence in Logistics and Distribution

This paper addresses the disruptive nature of Artificial Intelligence (AI) for logistics and distribution sector (Aipoh et al., 2025). Artificial intelligence technology is also increasing efficiency, reduced operational expenses, and improved last mile experiences by agile route planning, autonomous trucking, and drone delivery networks. Machine learning systems evaluate contextual traffic, weather, and road impositions such that routes for deliveries are optimized to reduce fuel usage and reduce downtime for operations. Share use of autonomous delivery is expanding, as transport vehicles of automation like self-driving trucks or drones, are changing last-mile and long-haul transport and increasing speed and reliability while reducing the need for human involvement. AI systems use different criteria like computer vision, sensors and prediction analytics in making real-time decisions and wayfinding as part of the operational process. Although there are called for benefits, regulatory issues with compliance, safety, and public confidence need to be addressed prior to widespread use. This paper provides a comprehensive review of the literature concerning logistics applications of AI and a discussion on the future of service implications of AI adoption.

Harnessing Drones for Faster, Cheaper, and Greener Logistic Solutions in Challenging Environments

This thesis investigates the potential of unmanned aerial vehicles (UAVs), or drones, in transforming global supply chains for improved sustainability and efficiency. The study examines the integration of drones within the logistics industry, focusing on risk management, sustainability, economic factors, and social reliability. Through case studies and data analysis, the research highlights the advantages of drones in reducing delivery time, minimizing carbon emissions, and optimizing last-mile delivery. (Iqab, 2024) investigated the potential of AI-powered drones for faster, cheaper, and greener logistic solutions, particularly in challenging operational environments. The thesis explores innovative materials and technologies that enable drones to withstand harsh weather conditions, drawing inspiration from aerospace advancements. A key proposed solution involves the collaboration of humanoid robots with cargo drones to facilitate door-to-door deliveries while ensuring safety, privacy, and security. The study concludes that while drones have the potential to revolutionize logistics, regulatory and technological barriers must be addressed to maximize their benefits.

Potential Role of Technology Innovation in Transformation of Sustainable Food Systems:

This review explores the impact of technological innovation on sustainable food systems (SFSs), focusing on advanced digital and agricultural technologies. The analysis demonstrates clearly ongoing and increasing challenges to global food systems from the double challenged of population growth on the one hand, climate change and resource depletion on the other hand. The authors considered several emerging technologies that may help in moving food production, processing, and disposal from current methods to alternative means, including IoT, AI, machine learning, drones, robotics, and gene editing. Several action initiatives are identified by the authors. These include, but are not limited to, changes in policy, market and financial incentives, social license to operate, and funding stability to accelerate technology use and uptake. The authors conclude by recognizing that technology has significant potential to improve food security and sustainability, but regulatory regime and the trust of stakeholders will be essential to provide a pathway for action (Khayyam et al., 2021).

Sensing, Smart and Sustainable Technologies for Agri-Food 4.0

The article aims at discussing the role that smart, new and sustainable (S3) technologies play participating in the development of Agri-Food 4.0, the current age of digitalization, automation, and efficient resource use in agriculture. The piece discusses the ways through which emerging technologies based on Internet of Things (IoT), artificial intelligence, robots, and automated systems are being adapted in the agri-food production process to increase productivity, sustainability, and profitability. The article suggests the S3 idea as a structure for the creation of new technologies for tackling problems of climate change and food safety and market competitiveness.

The article provides illustrative examples of cases of S3 technology being emerging, like, smart greenhouses, farming drones, and hexapod robots for crop monitoring. It makes a case that there is a requirement for policy change in a bid to introduce Agri-Food 4.0 technologies, sector standards should be established, and research investments and development should be undertaken for the realization of Agri-Food 4.0 potential.

Identifying Technological Enablers and Potential Future Applications in the Food Sector

This article discusses the transition from Food Industry 4.0 toward Food Industry 5.0, emphasizing the transition from solely technology-driven growths to a more human-oriented, sustainable, and flexible method. While Food Industry 4.0 emphasized digitalization, automation, and Data-driven efficiencies, Food Industry 5.0 connects human intelligence and emerging sustainability and responsible production techniques. The article recognizes key technical enablers like regenerative artificial intelligence, Internet of Everything (IoE), cyber-physical systems, 4D printing, redactable blockchain, edge computing twins. The future of mini and nano sensors was also included in the article, joint robots (cobots), intelligent drones, and 6G networks in revolutionizing the culinary industry of the future. The report argues that Industry 5.0 is not a revolution in itself but an evolution of industry 4.0, aimed at building more flexible and more resistant food supply chains (Carmona-Torres et al., 2024). The research places emphasis on the inclusiveness of new technologies while assuring moral considerations, environmental sustainability, and harmonious co-existence among human labor and machines.

Exploring the Role of Drones and UAVs in Logistics and Supply Chain Management: A Novel Text-Based Literature Review

The purpose of this paper was to rethink how the complementary and combined nature of drones and unmanned aerial vehicles, and the use of a new text-mining methodology to investigate the emergent nature of these items in logistics and supply chain (LSC). The authors shared their review of 5,364 papers written between 1978 and 2023 which provided them with ten unique perspectives on how drones fitted as part of LSC practices such as sustainable supply chains, scalability on a world-view platform, establishing mechanisms for pandemic response, and adoption of multimodal transport. As part of their manuscript, the authors explained how the implementation of the Latent Dirichlet Allocation (LDA) model would allow reader to classify the research landscape for future research directions - and glean the relevance of the future opportunities for drone-based applications as a mechanism to understand efficient delivery practices, warehouse automation, smart distribution, and occupational best practices for LSC all while developing knowledge about challenges such as regulations, fear of the security issue, and technical hurdles (Jahani et al., 2024). Overall, the authors derived conclusions about how there will be future opportunities for research, and also provided useful insights for future groups of researchers to consider ways to engage policy and technology so it can develop realization of the advantages of drones in future supply chains moving forward.

The Future Food Chain: Digitization as an Enabler of Society 5.0

This study explores the role of digitization in transforming food supply chains (FSCs) to align with Society 5.0, a human-centric and technology-driven vision for future societies. The paper highlights how globalization has expanded food availability and choice while also increasing complexity and reducing transparency in FSCs. Issues such as food fraud, safety, and security concerns have amplified distrust among consumers. The study emphasizes the importance of emerging digital technologies, including blockchain, artificial intelligence, IoT, and big data, in improving traceability, sustainability, and resilience in the food sector. The research connects FSC digitization with the UN Sustainable Development Goals (SDGs), illustrating how digital transformation can promote food security, safety, and nutrition (Keogh et al., 2020). The paper argues that multi-stakeholder collaboration, policy development, and digital integration are essential for achieving the goals of Society 5.0 and ensuring a sustainable global food system.

Building a Resilient, Sustainable, and Healthier Food Supply Through Innovation and Technology

Jozef L. Kokini, Mary Ann Lila, Anne S. Meyer, & Liangli Yu (2021) highlight how this study addresses the problems in today's food supply chain. The authors argue that the world needs innovative ways to produce, process, and distribute food. With an ever-changing world (in particular population growth; increasing diet-related chronic disease; and climate change), the case is made for innovation in the agri-food industry for long term goals of food system security, sustainability and health. Therefore, the paper was written to present certain technology developments that can be viewed as potential solving pathways, including, artificial intelligence, robots & automation, nanotechnology, gene editing and plant-based proteins. The authors also highlight that personalized nutrition, digital food tracking, and smart packaging are avenues for advancing sustainability in the agri-food system, and food safety improvements by reducing food losses. Although technologies can help in developing food systems, the authors note that eventually we will need to enable and continue to support equitable access and participation to equity and efficiency across the food system.

Role of Artificial Intelligence in Achieving Global Food Security:

The article explores different opportunities for using artificial intelligence (AI) to combat the global food security quandary created by climate change, lessening natural resources, and the ineffectiveness of global food systems. A key concept of global food security is the Food and Agriculture Organization's (FAO) four pillars of food security: food availability, food accessibility, food utilization, and food stability. Through examples of AI-based innovations the authors draw attention to AI-specific inputs and solutions such as machine learning methods for crop yield predictions, autonomous systems for irrigation, and smart controls in food supply chain management), to illustrate how they are useful in reducing food waste and improving the quality and efficiency of food redistribution (Chamara et al., 2020).

Workflow

Figure 3. Flowchart of data processing method of drones

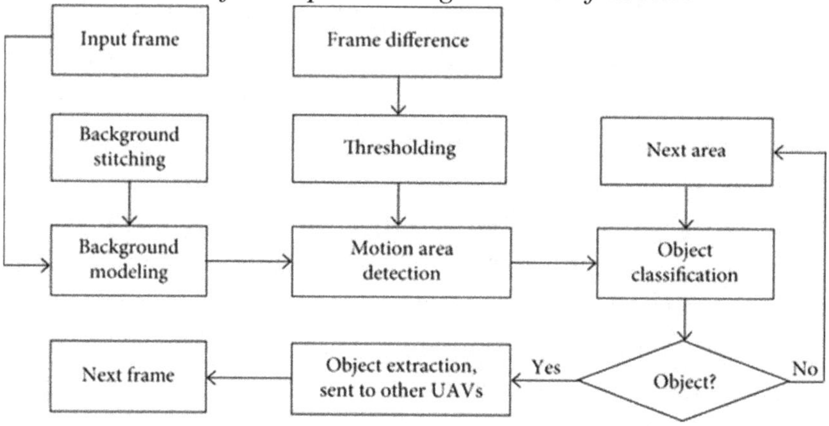

The AI Food Revolution: Reshaping Food Sciences through Artificial Intelligence

This dissertation will consider the spectrum of possibilities whereby artificial intelligence (AI) is influencing food sciences, particularly, agriculture, food processing, new product development, and consumer behaviour. AI projects that utilize machine learning, predictive analysis, and intelligent automated decision-making capabilities has the potential to revolutionise how farmers control their crops leading to improved agricultural productivity, or quality control. The dissertation will provide examples of AI as it relates to precision farming, one of the many potential areas where farmers are looking to improve yield intermediary to crop input and wasted resources. AI is also altering food processing - food safety feature to indicate potential foreign materials, and possibly, reduce, and if possible, eliminate batch-to-batch variability. The dissertation will be considering AI influence for personalised nutrition, whereby a data set provided adequate detail to predict - not dictate - human food selection behaviour. The spoils of conversation thereby brought a number of concerns related to ethical matters, such as: data privacy; risk of homeostatics bias on algorithm decisions; Design for operational support; and most significantly, ethical collusion; and the conclusion of the verbal quotes from experts will in the end conclude with a need for code and, regulations, (Ahmadi, 2023).

Internet of Things-Enabled Food and Plant Sensors to Empower Sustainability

This research review completed by Azahar Ali (2023) suggests the opportunity to leverage new, innovative, and IoT-enabled sensors with AI to improve food safety and monitor plant health and sustainability in agriculture (Kachouei et al., 2023). Wearable microfabricated sensors with IoT, which can provide real-time status on many aspects of food or plants including quality, contaminants, toxins, and plant growth, will improve predictive analysis and decision-making. The research review discussed some of the newer sensor technologies such as: 3D printing, nanomanufacturing, and wireless sensor networks to improve food traceability, improve crop management, and reduce the impact on the environment. The review touched on challenges such as scaling up, compliance to regulations, and data security, which could slow the spread of IoT-based food and plant sensors. The author suggested future research directions about AI-driven risk assessment models and the design of cheaper, remote sensing options to promote smart and sustainable agriculture

A Brief Review of Artificial Intelligence Robotics in the Food Industry

Hwa and Chuan (2024) evaluate the influence of Artificial Intelligence Robotics (AI Robotic) on reshaping the food sector while recognizing challenges such as labour shortages, quality control, and food safety. Automation has influenced food manufacturing, processing, and packing through vision technology and AI-driven robotic systems, improving operational efficiency in producing, transferring, storing, and packing food. This study also appraises advanced manufacturing technologies—such as autonomous robotic systems, cyber-physical systems, additive manufacturing, and cloud-based manufacturing systems—designed to reduce contamination occurrences. AI-driven robotics has demonstrated significant effects in food service automation, for instance in fruit sorting, poultry operations, intelligent logistics, transportation, and food service environments. The authors further reflect on AI robotics' impact on the food supply and delivery chains regarding hygiene and safety in the post-Covid environment. They conclude that AI robotic automation is a key component of Industry 4.0 and will continue to revolutionize the food industry by improving productivity, safety, and sustainability (Hwa & Chuan, 2024).

Significance of Drone Technology for Achievement of the United Nations Sustainable Development Goals

This study recognizes the transformative potential in drone technology to facilitate the attainment of the United Nations Sustainable Development Goals (SDGs) (Kitonsa & Kruglikov, 2018). Initially created for military suites, Unmanned Aerial (Unmanned Aerial Vehicles), commonly referred to as drones, have invaded every type of commercial, industrial, and social sectors such as police, roads, health, agriculture, and Environmental Conservation. This paper illustrates the role of drones in promoting sustainability by reducing carbon footprints, maximizing outputs from available resources, and providing low-cost solutions for problems such as famine, poverty, and disease control. The paper recognizes the importance of drones with respect to precision farming as they are used to track crops, apply pesticide accurately, and deliver goods from the farm to market. Drones are also seeing a growing use in health care delivery, especially to rural areas in developing nations.

Artificial Intelligence in Modern Consumer-Driven Food Systems

This study investigates the transformative influence Artificial Intelligence (AI) is having in contemporary food systems, specifically from farm to fork (Min et al., 2024). This article presents an AI-enabled Food System (AIFS) framework, which represents the confluence of AI technologies including computer vision, speech processing, machine learning, and their use in internet of things. AI determines personal food preferences, influences production farm, and provides a cyclical "consumer driven farm to table" process. The article discusses the use of AI in precision agriculture, food processing, chains of optimisations, and tailored nutrition. In addition, it discusses subjects such as AI governance, ethics, privacy, and interoperability. We believe that AI, when coupled with consumer abstracted and sustainable food policy, has the ability to make the world food strategy a more responsive and flexible system.

The Food Production–Consumption Chain: Fighting Food Insecurity, Loss, and Waste with Technology

The purpose of this study identifies how technology can be used to combat food insecurity and minimize food loss and waste, throughout the entire food production - consumption process (Grewal et al., 2024). The authors point to the value added by more advanced technologies including, artificial intelligence (AI), blockchain, UAVs, IoT, and cloud computing in improving food production, distribution, and waste. The study refers to food supply chain ineffectiveness, noting that nearly one third of all food produced is lost, while millions are undernourished. A proposed typology classifies technology's contribution to growing food production and enhancing food redistribution. The study presents practical examples like AI-based crop monitoring systems, blockchain-based food traceability, and intelligent food wastage management apps. From the study, we find that though technological advances can significantly improve global food security, regulation challenges and adoption by stakeholders remain crucial to achieving the United Nations Sustainable Development Goals (SDGs) for zero hunger and sustainable food production.

A Sustainable Blockchain Framework for the Halal Food Supply Chain: Lessons from Malaysia

This study responds to the question of whether blockchain technology can enhance transparency, traceability, supply chain efficiency in the halal food supply chain (2021). Owing to the importance of halal certification in the global food market, blockchain is introduced as a solution to prevent fraud, contamination, as well as mislabeling while boosting confidence between clients. The paper identifies five key issues that are faced by small to medium enterprises (SMEs) in adopting blockchain technology: complexity, cost, change management, sustainable production, alongside regulatory compliance. Reference is made to practical halal food Supply chains and provide a blockchain platform that is customized to SME needs. The study points out that whereas blockchain is beneficial, regulatory convergence, supply chain integration, as well as industrial implementation, are needed for its success. The findings help in developing the body of knowledge on the use of blockchain towards preserving food integrity, specifically in halal industries.

Analyzing Human Robotic Interaction in the Food Industry

This is a writing on increasing use of food supply chain robots in companies. From agriculture to packing and distribution (Grobbelaar, Verma, & Shukla, 2021). As safety norms and consumer demands for enhanced efficiency, robotics has gained highest significance in modern food production. Automation is described in the paper as positive in relation to hygiene, uniformity, less human error, specifically in packing as well as in lines that repeat. The use of robots in food has still not gained momentum like in other sectors Due to concerns regarding layoff losses as well as startup capital costs. Joint robots (cobots) are proposed as the solution in that these robots work with human workers to increase efficiency while reduction in manual interventions. The paper also considers advances in AI-based robotics such as smart vision, voice recognition, and multi-tasking capacities, which are enabling robots to adapt more in food manufacturing environments. Despite concerns about workforce displacement, the study concludes that robotics will redefine food production and create new opportunities for human workers in supervisory and high-skill roles.

AI-Enabled Waste Management for Sustainable Sewage Treatment

This article discusses a novel approach to monitoring waste and sewage pollutants using an AI-enabled waste management system (Suresh Babu,

Abirami, et al., 2023). Specifically, it proposes dividing sewage or waste into solid and liquid streams, which are then treated using resource recovery techniques to obtain distilled (pure) water and recover compostable material. Unlike previous works, this article estimates the additional waste that could be removed by breaking down waste management tasks into smaller zero-waste scenarios, creating products such as distilled water, clean compost, and organic fertilizer through nine types of AI-powered waste management systems. The potential for AI to provide real-time monitoring data enables accurate water quality assessment, problem investigation, and greater transparency with stakeholders and the public. The paper highlights that globally, approximately 80% of wastewater is discharged untreated or improperly treated, posing severe ecosystem and public health risks. It focuses particularly on un-managed sewage streams in India, where about 70% of sewage is unmanaged due to infrastructure failures or mismanagement, causing environmental pollution, groundwater depletion, and damage to aquatic biodiversity. The study advocates for AI-driven mitigation strategies to eliminate waste pollution, reduce contamination, and promote sustainable circular water resource management. It concludes that without AI strategies, waste management will struggle to advance beyond current limitations towards integrated, environmentally-friendly technological interventions (Suresh Babu, Abirami, et al., 2023).

Trustworthy AI for IoT Sensory Data in Agriculture: Enhancing Reliability and Security

This study is concerned with the integration of Artificial Intelligence (AI) with the Internet of Things (IoT) in farming, as importantly stressing the critical need to enable trusted AI for best-of-breed farm practices. It demonstrates how AI is leveraging the available IoT sensor streams from your farm equipment. including soil moisture sensors, weather stations, and drones to enhance farmers' decision-making. The writers observe that crop health, presence of pests, as well as environment information can improve efficiency and resource allocation to obtain high crop production. The research also presents solutions to the challenges of implementing credible AI, emphasizing transparency, accountability, and trustworthiness in AI-driven decision-making processes. Security concerns about data privacy, backdoor attacks, and AI biases are addressed to improve precision farming reliability. Ultimately, the research argues that exploiting reliable AI is dependent on secure working methods as well as robust ethical frameworks that define trust boundaries, in order for AI to become an unbeatable partner in agriculture's sustainable future (Suresh Babu, Koti Reddy, et al., 2024).

Comparison of AI-Based Drone Applications in Food Logistics

Numerous research works have investigated the essence of AI-enabled drone in food logistics, stating that these make the process faster, cheaper, and greener.

Optimization of Routes and Last-Mile Delivery

Aipoh et al. (2025) explored AI drone delivery systems through studying machine-learning algorithms in order to customize a delivery route depending on traffic as well as weather. Once again, in 2024 Iqab announced more benefits, declaring drones utilization to reduce transportation expenses and carbon footprints while achieving speed and last-mile rural and urban delivery efficiency.

Supply Chain and Inventory Management

Jahani et al. (2024) discussed integrating drones into supply chains as well as speculated that drone technology can benefit warehouse management operations such as stock-taking, scanning barcodes, and tracking in real-time.

Sustainability and Environmental Benefits

Kapoor et al. (2023) and Khayyam et al. (2021) had an overview of how AI-driven drones assist in sustainable food logistics by minimizing food waste as well as lowering the carbon footprints from conventional sources of transport.

Regulatory and Technological Challenges

On the contrary, research papers that have been published by Kitonsa & Kruglikov (2018) documents the regulations and challenges facing drone technology, cyber security and embedment into current-day logistics system.

METHODOLOGY

Research Approach: Qualitative Analysis of AI-Powered Drones

This study takes a qualitative route to investigate the contribution of AI-powered drones to food production and supply chain logistics. A qualitative approach is selected to develop an integrated and descriptive understanding of technological advancements, applications, and effects of AI-powered drones in the food industry (Creswell & Poth, 2018). Through the use of case studies, expert perspectives, and business reports, the research seeks to understand how AI-powered drones improve efficiency, sustainability, and productivity in precision farming, food processing, and supply chain optimization. The study explores AI-drone technology using a combination of scholarly articles, industry reports, technology companies, and agricultural associations to consider AI-drone technology's present state, and future directions (Patel et al., 2022). The authors use qualitative content analysis to examine the adoption trends and patterns of AI drones, impacts on operational efficiencies, and how this technology supports sustainable food production practices, (Creswell & Poth, 2018; Patel et al., 2022). Qualitative research differs from quantitative research because it excludes aspects of numerical and statistical information, providing a more in-depth examination and conversation. While quantitative research tends to present aspects of the technological, economic, and environmental implications of AI drones, the qualitative nature of this study allows for an overall succinct description of how the AI drones are making changes within the food sector. The authors provide valuable information for researchers, industry practitioners, and policy makers to consider regarding innovations related to food production and distribution.

Case Study Selection and Analysis

This study utilizes a case study approach to examine the real-world applications and effects of AI-equipped drones in food production and supply chain operations. The chosen case studies target companies, agricultural projects, and supply chain technologies that have effectively integrated AI-equipped drones to enhance efficiency, sustainability, and food security (Yin, 2018). Through an exploration of actual implementations, the research yields useful observations on the functioning, challenges, and potential for growth of AI-based drone technology in the food industry. Case study selection depends on their application to agriculture and food logistics, technology innovation, and quantifiable effects on farm yield, resource usage, and supply chain performance (Patel et al., 2022). Such case studies involve precision farming initiatives in which drones powered by AI are employed to survey soil conditions, check the health of crops, and automatically apply pesticides. Other case studies involving the food processing and distribution sectors illustrate how drones aid in quality control, reduce food loss, and enhance transport efficiency. The study contrasts information from industry reports, academic research, company documents, and expert interviews to provide a thorough analysis. (Yin, 2018; Patel et al., 2022) All case studies are measured based on critical performance indicators like cost reduction, productivity increase, environmental impacts, and scalability of AI drone integration.

Case Studies on Failed Implementations or Challenges

Regulatory Delays of Amazon Prime Air

Problem. As a result of the promise of 30-Minute deliveries by Drains, years of policy difficulty has since been harbored in Amazon Prime Air in the U.S. and Europe. The Federal Aviation Administration has imposed very strict limits on operations.

Consequences. Roll out is limited to selected test areas, but now it isn't clear whether this would be the case just reading based on changing laws and safety concerns.

Drone Delivery of UPS & Matternet in North Carolina

Issue. In addition to that, UPS has partnered with Matternet to start up food and medication deliveries with drones. However, repeated system failures and flight restrictions have forced a catastrophic delay in operations.

Effect. FAA certification was obtained; however, more operating costs and a huge chunk of population skepticism slowed down the expansion of the drone logistics to the urban areas.

Wing Delivery Failures in Australia

Issue. Delivery of food in Canberra, Australia, attempted by Alphabet-owned company Wing, has been met with complaints of noise pollution from the residents, hence regulatory scrutiny. Weather-related incidences are also tied to strong winds that drones were not able to fly through.

Severity of Impact. The terms had entered into place in such a fashion that the Regulators set the stage for far too strict flight with conditions that hardly rendered large scale operating of drones unbelievably feasible.

Data Sources: Industry Reports, Expert Interviews, and Literature Review

The study employs various sources of data, such as industry reports, expert interviews, and a literature review, in order to provide the most holistic and balanced analysis construed from some of the available sources of data. These data sources provide reliable accounts of the AI technology and AI drone current state, issues with technology, and what the future holds in respect to food production and logistics (Yin, 2018). Industry reports are also highly reliable and useful data sources that offer current understanding of AI Drone application in agriculture, food processing, and supply chain management. The reports from organizations such the Food and Agriculture Organization (FAO and) and the World Economic Forum (WEF), and leading agri-tech companies were also used to glean market trends, actual or estimated, the uptake of AI Drones in food systems and technology advancements. These reports also provide some statistical information on productivity and efficiency, cost savings to businesses and environmental benefits of using an AI Drone. Interviews with experts provide first-hand information on the actual imperatives and advantages of AI application in food production (Patel et al., 2022.Views of agricultural scientists, technology developers, and logistics specialists assist in confirming secondary source findings while providing context-specific uses of AI-driven drone technology. Literature review entails the scrutiny of scholarly research articles, journal papers, and conference papers on AI in farming, drone technology, and automated food logistics. Through the exploration of prior work, this study establishes major technology trends, conceptual frameworks, and gaps in available knowledge. Literature review also guides analysis of the history of AI-based drones and the long-term impact of their sustainability on the food sector. By combining these three sources of data, this study provides an even-handed and evidence-based evaluation of AI drones, emphasizing their revolutionary impact on food production and supply chain efficiency.

AI and Drone Timeline in Food Logistics

Early 2000s

The logistics and food processing industries started adopting rudimentary automation. At this point, AI was mostly concerned with warehouse management and quality assurance.

2010-2015

This decade witnessed AI technologies leading the way for machine learning-based forecasting systems, especially in weather-related supply chain management. Drones were primarily used for surveillance of agricultural activities during this time.

2016-2019

We saw the advent of AI-driven vision systems in drones, which were meant for real-time food quality monitoring and automated warehouse inspections.

2020-2023

AI-powered drones revolutionized food delivery apps, transforming last-mile logistics in a big way with a sustainable approach. Attempts were also being made to utilize blockchain technology to enhance food traceability.

2024-Present

Research continues to examine the possibilities of using AI-powered swarm drones and edge computing to enable real-time decision-making for food logistics while maintaining energy efficiency and regulatory compliance.

CONTRIBUTION TO THE FIELD

Advancing Knowledge on AI Drones in the Food Industry

This study adds to the current literature on AI-drone technology through a consideration of the potential of AI-driven drones in food production and supply chain operations. Through examination of current practices, technology advancements, and business trends, the research gives the reader an insight into how AI drones make food production, delivery, and storage more efficient, sustainable, and productive (Patel et al., 2022). These sources provide valuable information on the context of AI drones (lessons learned), challenges that need to be addressed, and potential opportunities for food production and distribution (e.g., warehousing, trucking) (Yin, 2018) related to AI drones. Other sources of value are industry reports to find the latest levels of drone usage for agriculture, food processing, and supply chain management. For example, information from the Food and Agriculture Organization (FAO), World Economic Forum (WEF), and they will also describe current market dynamics, rates of adoption for the drones and advancements in technology by top agri-tech companies. These white papers and reports will also have statistical evidence for efficiency, savings, and positive environmental outcomes of using AI-controlled drones. Interviews with experts across all the fields, e.g., experts in precision agriculture, drone manufacturers, AI developers, experts in any part of the food supply chain. Expert interviews provide firsthand data that support the practice imperatives and possible benefits for AI implementation into food (and products) production (Patel et al., 2022).

Addressing Challenges in Drone Adoption and Integration

One of the main deterrents is regulatory constraints, since aviation and food regulation is different in different nations, rendering drone operations incompatible across the board. Keeping up with changing regulations for drones, airspace controls, and food regulations necessitates continuous updates and cross-industry cooperation. Financial limitations are another major obstacle, whereby the initial setup costs of AI-based drones, combined with the cost of maintenance and training for operators, are prohibitively expensive for small and medium-sized businesses (SMEs) (Patel et al., 2022). Technological limitations like short battery life, low payload capacities, and poor connectivity in remote farming regions influence drone efficiency and real-time data delivery. The deployment of AI-powered drones into current agricultural and supply chain networks needs to be supported by strong data infrastructure and advanced technical skills, which are beyond the capabilities of most farmers and companies, (Patel et al., 2022). Privacy and data protection issues also exist since drones gather, store, and transfer confidential agricultural and supply chain data. Illicit access or misuse of this information threatens food production systems and supply chain openness. To overcome these issues, there is a need for supportive policies, technological innovation, and industry cooperation. By enhancing affordability, boosting AI capacity, and providing clear regulatory guidelines, the food industry can best utilize AI-powered drones for greater efficiency, sustainability, and food security

Providing Insights for Researchers and Industry Practitioner

This research is valuable to researchers and industry practitioners alike by detailing the applications of AI-powered drones in food production and supply chains, their benefits, and their limitations. For researchers, this research establishes a foundation for vehicles to study the impact of automation using AI technologies, food production, precision farming, and sustainable food chains, (domestic and globally). It outlines contemporary technological advances and literature gaps and motivates future work towards making drones more efficient, integrating AI and the Internet of Things (IoT) with blockchain technology, and eliminating regulations, (Patel et al., 2022). For businesspeople, (e.g., farmers, agri-businesses, and supply chain managers,) this research is a working manual for adopting AI-powered drones for their operations. In addition, it provides best practices and case studies on the use of drone technologies for crop surveillance, efficient resource management, and last-mile delivery services as they are used in agriculture. It also details solutions to the adoption of drones such as cost limitations, regulation issues, and employee training to operate AI powered drones. Importantly, this study connects research done at the academic level with the current issues facing industry, revealing the opportunities for beneficial evidence-based decisions and sustaining innovation and efficiency in food production systems, (Patel et al., 2022). AI-powered drones can change agriculture disrupting food systems.

Problems and Shortcomings of Previous Studies

Regulatory Constraints

- It differs by sections that restricts the large-scale deployment of drones.
- Airspace regulation and privacy issues restrict drones in logistics expansion.

Technical Limitations

- Short battery life and low payload capacity render the larger quantity food transport by the drones impossible.
- It is not yet possible for AI to identify the decision processes in complex urban areas.

Data Security and Integration

- Blockchain integration for food traceability is in its early stages (Keogh et al., 2020).
- Cybersecurity concerns related to AI-based drone system require closer examination, (Keogh et al., 2020).

Economic and Social Aspects

- Expensive installation and maintenance prices keep small firms away from aerial logistics waters (Patel et al., 2022).
- Public acceptability and labor market implications due to AI-powered automation need consideration as well.

Regulatory Frameworks Governing AI in Logistics

Policies governing AI-driven logistics vary across regions but share common objectives: enhancing efficiency while ensuring safety, compliance, and ethical use.

- **United States:** The Federal Aviation Administration (FAA) imposes strict regulations on autonomous drones for commercial delivery, requiring Part 135 certification for beyond-visual-line-of-sight (BVLOS) operations. Amazon Prime Air, for instance, struggled with FAA approval, delaying its drone delivery expansion.
- **European Union:** The AI Act (expected 2025) classifies AI applications into risk categories, with autonomous logistics falling under "high-risk" due to safety and privacy concerns. The EU also enforces General Data Protection Regulation (GDPR), impacting AI-driven data collection in supply chains.
- **China:** The Artificial Intelligence Regulations mandate real-time monitoring of autonomous logistics systems, requiring black-box data recording for accident investigations. The rapid expansion of JD.com's drone delivery system is subject to these controls.

Challenges and Future Outlook

Regulatory and Legal Challenges

The regulatory landscape for AI-powered drones is a key factor in how quickly and effectively these technologies can be used in the food industry Drone flying in the USA categorically adhere to Federal Aviation Administration's (FAA) Part 107 Small UAS Rule. This rule places limitation on maximum altitude, needs line-of-sight visibility, needs operator certification, and establishes zones of operation. Long-standing annexures, such as the Remote Identification (Remote ID) requirement will force drones to broadcast their ID and location. This facilitates authorities to verify compliance and offer increased airspace safety (FAA, 2023). Whereas while these frameworks enable commercial use of drones, the American regime still presents obstacles to beyond-visual-line-of-sight (BVLOS) operations, night flights, or fully autonomous missions. They all opt for waivers or very restricted authorizations. What does the European Union ask for? A single, risk-centered approach to EASA Regulations (EU) 2019/947 and 2019/945 system, drone flight is in one of open, specific, or certified categories. Each category has separate needs for pilot abilities, technical capability, and operation authorisations. The EU's "U-space" concept puts drones in a regulated traffic environment as from the very beginning, streamlining operating clearance for large-scale or complex missions (EASA, 2024). For global food-industry application of drones, these regulatory differences induce high implementation cost. A drone delivery approved under FAA regulations can integrate technical changes, software upgrades, or reconfigurations of operation in order to comply with EASA standards. Secondly, in the EU, data governance also overlaps with aviation law since GDPR has stringent treatment protocols of photos or sensor data that hold personal information. Therefore, any extensive rollout of scales needs to incorporate regulatory compliance and regulatory intelligence into the previous stages of research and development in order to scale across jurisdictions (Keogh et al., 2020; Carmona-Torres et al., 2024).

Cybersecurity and Data Privacy

As AI-powered drones become more autonomous and connected, cybersecurity and data privacy become crucial concerns. The drones, their communication channels, ground stations, and cloud analytics platforms all have potential vulnerabilities. Threats include unauthorized access to live video feeds and malicious command injection that could disrupt operations or lead to crashes (Suresh Babu, Koti Reddy, et al., 2024). The risks are especially high in the food industry, where drones may manage sensitive logistics data, delivery routes, or proprietary agricultural monitoring images. In the U.S., agencies like the FAA, CISA, and DHS are working to integrate cybersecurity principles into unmanned aircraft system (UAS) policy. The FAA's cybersecurity strategy follows NIST standards and stresses the importance of secure hardware and software design, secure boot processes, signed firmware, and encrypted communications (CISA, 2023; NIST, 2023). Meanwhile, CISA offers practical advice on vulnerability assessment, patch management, and incident response for UAS operators. In Europe, EASA and national cybersecurity agencies match drone cybersecurity requirements with the EU Cybersecurity Act and GDPR. The latter adds complexity: even aerial images of farmland can be considered personal data if they identify individuals or private property.

GDPR compliance requires clear data processing purposes, storage limits, and ways for individuals to exercise their data rights (Keogh et al., 2020).

Ethical Issues and Public Perception

Public perception of drones in food delivery, warehouse automation, or agricultural monitoring depends on ethical considerations as much as on technical performance. While stakeholders often point out the benefits, such as faster deliveries, reduced emissions, and precision agriculture, communities may express valid concerns about privacy, safety, and noise pollution (Iqab, 2024). For instance, delivery drones flying over residential neighbourhoods could unintentionally capture images of private backyards or be seen as intrusive, even if they don't store any data. These concerns contribute to a larger ethical discussion about balancing technological progress with individual rights and community well-being (Hwa & Chuan, 2024). Privacy by design is crucial. These unmanned aerial systems should also not collect anything without personal information, automatically safeguard personally identifiable data, and operate only in designated airways. Transparency about what data is collected, how developed, where and who has access will help create trust. Perceptions are influence by safety as well; low incidence of drone-crash stories can negatively impact confidence in the drone. Safety features, such as redundancies in navigation, automatic collision avoidance, and well laid out emergency landing protocols, along with reliability improvements also provide citizens with some comfort (Jahani et al.., 2024.

Issues, Problems, and Trends

Although AI-enabled drones offer great potential in food systems, there are definitely obstacles to overcome. In addition to knowing the cyber vulnerabilities of autonomous devices and vulnerability of a drone's operation not only within a network and multiple AI algorithms but from hackable functions, their integration into food systems raises some hurdles. As some researchers indicate, the perception of drones could be that it is bureaucratic or noisy or not safe and that perception will firmly limit the acceptance of their use in an aggregation operation context. Regulation will vary by region; the FAA in the United States has operational limits under Part 107, whereas the European Union has created a classification system in the European Union Aviation Safety Agency (EASA) Risk Classification. Other countries such as China and India are still testing the ground and developing prospective guidelines. There is much academic research focusing on developing secure AI frameworks, strengthening transparency of data handling, and creating global policies harmoniously. Addressing ethical and regulatory aspects of these technologies and practices will be contingency to verifying drones became a trusted part of the sustainable food system.

Case Study

Success. In 2023, a major grocery chain in Finland used AI-assisted drones for last-mile deliveries in rural areas. They operated in harsh winter conditions. The fleet utilized thermal imaging for navigation and optimized routes in real time to reduce battery consumption (Keogh et al., 2020). This project cut delivery times by 40% and decreased carbon emissions by about 25% compared to deliveries using vans. Key factors in this success included regulatory compliance with EASA's "specific" category, a strong cold-weather hardware design, and community outreach that enhanced public acceptance (EASA, 2024).

Case Study

Failure. In 2022, a drone delivery trial in a Southeast Asian city faced suspension after frequent GPS signal loss in dense urban areas led to multiple aborted flights and two minor crashes (Iqab, 2024). Technical issues worsened due to insufficient operator training and unclear compensation policies for damaged property. This case underscored the need for feasibility studies tailored to specific locations, backup navigation systems, and insurance plans before scaling up.

Figure 4. Global use case map (world map showing Japan, Netherlands, USA, Kenya case stats)

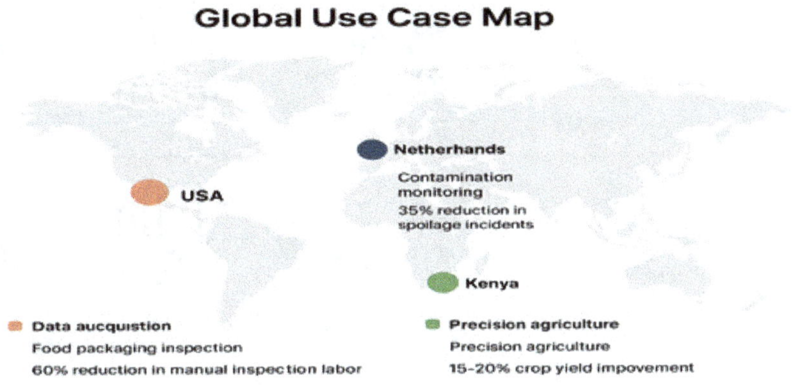

Climate and Economic Limitations

In very hot climates, battery performance suffers, which reduces flight duration and raises maintenance costs (Patel et al., 2022). In developing countries, high initial costs and the demand for skilled workers might limit adoption, even if drones present operational advantages (Kitonsa & Kruglikov, 2018). Gaps in infrastructure, such as unreliable internet service, can obstruct real-time AI analytics and decrease operational effectiveness. The trend remains hopeful. Progress in edge computing, battery technology, and regulatory developments is gradually addressing these issues. Nonetheless, recognizing and preparing for these challenges is essential to turn AI-powered drones from pilot programs into reliable, scalable tools in the food industry.

Figure 5. The use of drone technology

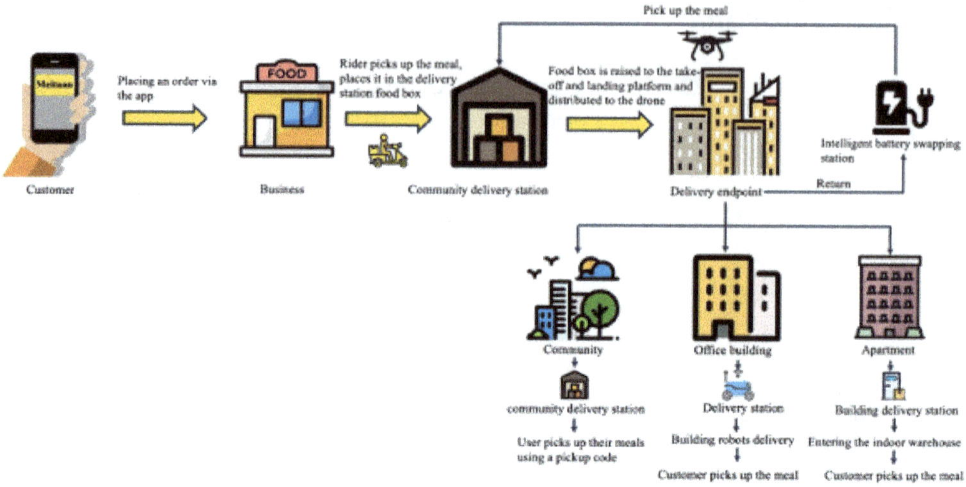

AI-powered drones are moving beyond field logistics and entering important areas of the food industry. New deployments demonstrate a wide range of applications:

Food Packaging Inspection. In Japan, drones equipped with AI-based computer vision inspect sealed food packages for defects, such as incomplete seals or damaged labels, before the products leave the factory floor (Miranda et al., 2019). This system cuts manual inspection work by 60% and improves consistency in quality control.

Contamination Monitoring. In large meat-processing plants, thermal and hyperspectral imaging drones identify contamination risks, such as mold growth or temperature issues in storage areas (Kapoor et al., 2023). AI algorithms highlight anomalies in real time, allowing for quick action before contaminated products reach distribution.

Warehouse Inventory Management. In U.S. cold storage facilities, autonomous drones linked with warehouse management systems scan barcodes and monitor pallet locations in refrigerated settings. This reduces inventory check time by 80%.

These examples show that AI drones are not just for farming logistics; they can also be used in processing and distribution points, improving efficiency and safety throughout the food supply chain.

Technical failures, such as GPS loss in urban areas or sensor calibration issues in extreme temperatures, can lead to operational downtime (Iqab, 2024).
Climate factors, like high humidity in tropical warehouses, can damage sensors and batteries more quickly (Patel et al., 2022).
Economic challenges exist in developing countries, where high initial costs and a lack of skilled workers slow down adoption (Kitonsa & Kruglikov, 2018).
By refocusing some research on processing and logistics, future studies can better address integration challenges. This can help move away from repeatedly discussing farming use cases and emphasize the wider strategic benefits of AI drones in the food industry.

ANALYSIS AND FINDINGS

Benefits of AI-Powered Drones in Food Production

Drone solutions and artificial intelligence will develop a more efficient, dependable, and sustainable food production system if food producers want to be more conscious and steward Sustainability and the Environment. One advantage of drone solutions, I don't know if I want to say advantage, but you know where I am going with that, is that it helps with precision farming. Consider drones, just like everything else, being embraced by artificial intelligence that incorporates sensors and imaging devices to measure and quantify soil health, detect weeds and diseases, and help determine the right amount of water to irrigate. Not only that, but help reduce waste, think about in water, pesticides, and fertilizers. Pesticides and fertilizers are applied when needed and calculated to be applied when only needed, therefore providing peace of mind for the producer that the application is appropriate, and will better yield a greater tonnage, and also, ultimately and most importantly, lower overall input costs to the producer. Drones can provide a great many additional value add's, for example; they can automate and help with planting, spraying, and harvesting therefore reducing human interaction or in most part human labour. The use of drones in food production systems promotes better food security, reduce operational costs and ultimately create more efficiency in the agricultural system.

Figure 6. Advantages of drones in food production

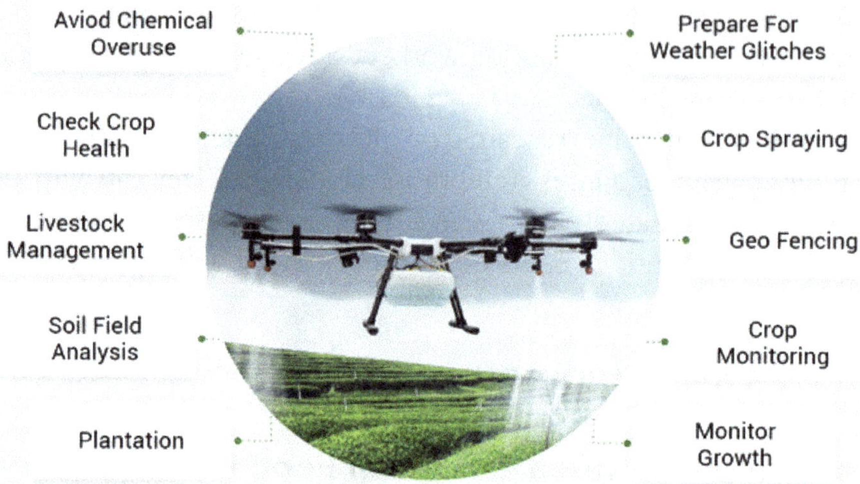

Efficiency and Cost Reduction Achieved with AI Drone

AI drones increase efficiency and therefore tend to lower expenses by cutting wastage and automatizing any procedure critical to food production as well as supply lines. The question of workforce effort alone generates larger benefit in terms of Labour process as they conduct manual schedules such as crop checking, application and assessment of pesticide treatment soil conditions, all without any human interaction, preserving huge quantities of labour time. The utilization of AI drones also minimizes human dependence, hence producing. Recoverable farm operation costs to farm players. Use of AI drones Would improve resource utilization through the distribution of accurate information to use water, pesticides, and the use of them in time, therefore reducing wastage, savings in cost on inputs, as well as redundantly augmenting crop production. From a logistic perspective, AI drones are used for streamline supply chain through good tracking to stocks, reducing food spoilage and quicker delivery while still achieving significant cost savings. In addition, AI drones support predictive and preventative food produce maintenance by detecting potential problems before they become an impending large-scale issue with excessive costs. Improving productivity, financial viability, and in the end, profitability using every worker focused, employ-efficient workflows possible usually with no operational cost. Studies have indicated drone logistics provide faster, cheaper, more green options, particularly in complex operational contexts (Iqbal, 2024).

Figure 7. Advantages of AI in drones technology

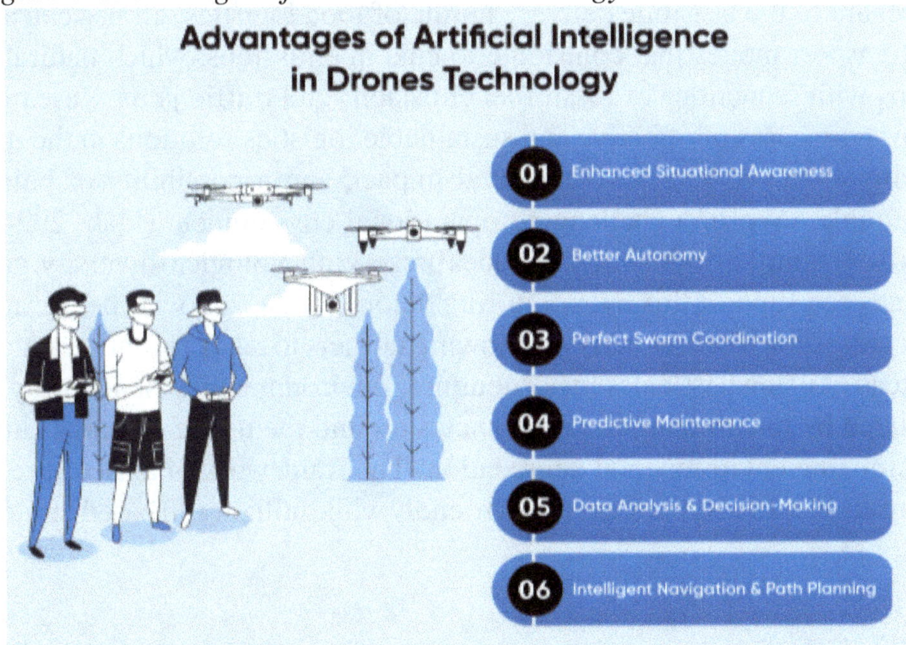

Enhancements in Sustainability and Environmental Impact

AI Drones are a significant contributor to environmental sustainability through reduced negative environmental impact of agriculture. AI Drones utilize precision agriculture, which helps conserve, and enhance the use of water, fertilizers, or pesticides by utilizing what is called "precise applications" with none or very little waste. The result is that there is no destructive flow into the ecosystem through the applications. Using precise applications in the areas where the water, fertilizers or pesticides are required and does result in less soil erosion and water contamination leading to environmentally friendly agriculture. Another important environmental benefit is the reduced carbon emissions that are realized. Conventional farming systems are highly dependent on fossil fuels, as farmers and farm workers rely on high energy machinery. While AI drones at times require some energy, they require very little energy, and it is possible to find both electric and solar-powered drones making the environmental impact remarkably reduced. In regard to the last-mile delivery hurdle of food logistics, drones can act as transport modes that contribute to cuts in emissions, which naturally helps with reductions of fossil fuel emissions and traffic jams. Research shows use cases of effective and sustainable logistics solutions in the use of drones to mitigate environmental impact, and a possibility of being profitable, even in a challenging operational environment (Iqab, 2024). Environmental sustainability includes preserving biological diversity, and drones can support this by monitoring deforestation, soil well-being, and detecting crop diseases early on allowing farmers to take preventive action instead of using pesticides in horticulture. Environmental sustainability is realized by maximizing the efficiencies around the use of resources and minimizing environmental degradation. The framework of agriculture is more efficient and environmentally friendly while utilizing AI based drones.

Figure 8. Global AI in drone technology market

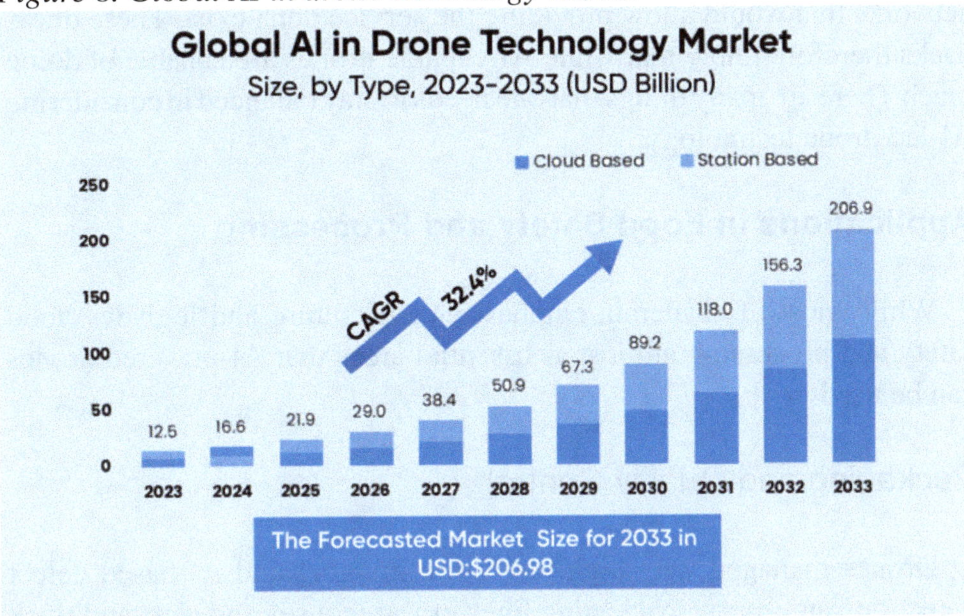

Risks & Limitations

While most of the conversation about AI-powered drones has centred on efficiency and sustainability, it is equally important to acknowledge the risks and shortfalls. A major concern will be the reliance on AI systems. The entire food supply chain operation could come to a standstill if algorithms fail due to bad training data, software bugs, or unforeseen circumstances. Unlike human workers, drones aren't sufficiently adaptable when things go awry.

Another limitation with AI-powered drones will be continuing costs and upkeep to use them. Bigger capacity drones are going to take continual drone battery changes, restore the sensors and equipment to working order, and will require experienced technicians to manage calibration and repairing. This will make it opportunistic for small or mid-sized companies in the food businesses as nothing can bring on cash more quickly than eliminating costs of poorly defined markets. We should talk about flexibility in rural areas and under-developed areas, where GPS mapping, internet capacity, and power supply are all very limited, and the drone networks that would allow providing the service don't exist. These drawbacks therefore imply that while AI-capable drones are capable of doing much There are many things that one needs to stay balanced in considering AI-led drone technology.

Applications in Food Safety and Processing

While most of academia emphasizes agriculture and logistics, food safety and processing are just as essential areas that AI-powered drones can be used in.

Packaging and Quality Control

Drones managed with high-resolution imaging and AI-based defect detection can observe packaging lines in warehouses and alert and track problems such as torn packaging, labelling issues, temperature violations in cold storage, etc., to ensure that consumers only receive safe products.

Contamination Monitoring

AI-based drones fitted with multispectral sensors can monitor for biological contaminants such as mold or bacteria on stored food products. With large-scale storage or distribution networks, the airborne monitoring capabilities provide continuous monitoring capabilities, which will minimize the chance of unsafe products contaminating the food supply.

Warehouse and Inventory Functions

In addition to safety, drones will assist in processing by identifying bar codes, identifying and cataloguing expiration dates, and mapping available warehouse space. As a result of successful documentation, the accuracy of inventory counts reduces but also ensures that secure products are handled with minimal risk of produce contamination with manual labour warehousing duties. Utilizing AI-powered drones is particularly important in large volume, inefficient environments such as frozen food warehouses.

These examples are equivalent in size and scope, and they demonstrate that AI-enabled drones are not limited to either the farm level nor the delivery of food. By supporting food packaging and food contamination, along with warehousing functions, drones can develop safer and reliable food distribution systems after harvest.

Analytical Insights

All of this empirical evidence adds weight to the AI-assisted drone argument - commercially speaking. For example, inventory management in warehouses from drones decrease stock-checking times by about 40%, compared to manual inspections; however, inventory management only describes a piece of an operational puzzle providing operational efficiencies. Similarly, Japanese pilot projects showed drone crop monitoring can improve yield prediction accuracy by upwards of 20–25%, allowing farmers to make informed pesticide and fertilizer usage decisions. In logistics adoption, Amazon's air trial (for Prime Air) suggested average delivery times of under 30 minutes, compared to traditional delivery services averaging 1–2 hours for short-distance orders.

FUTURE SCOPE FOR RESEARCH

Future Directions in AI-Powered Drone Applications for the Food Industry

The application of AI-based drones in the food sector has considerable potential for growth and advancement with automation, efficiency and sustainability. One research opportunity is to merge AI to reconsider emerging technologies (i.e. Internet of Things (IoT), blockchain, and edge computing) that would allow sharing in fast/real time, predictive analytics and better traceability in the food supply chain thereby achieving better food safety and quality standards. A separate research interest that has substantial potential would be related to autonomous swarm drone systems, whereby drones would work collectively and operate as one to conduct broadacre agricultural related functions (i.e. precise plantings, pest co-management, agriculture (Patel et al., 2022)). with minimal human intervention. Whatever the case, AI drone adaptability issues should also be removed so that the machine learning models will be able to improve the adaptive capacity for the drone to make decisions autonomously in complex, rapidly evolving farm contexts. Future research will also investigate energy-efficient drone technologies (i.e. solar powered drones) and battery systems for even more sustainability.

Integration with Blockchain for Traceability and Transparency

The utilization of AI-based drones combined or linked with blockchain has the potential to enhance transparency in the food sector. Blockchain is a distributed, tamper evident, digital ledger that may enable the secure storage of your AI drone information in real-time to ensure accuracy in the food supply chain. Drones gather and process vital information related to Crop health states, confirmed pesticide uses, and harvesting dates and times which can then become records that create more transparency in food safety and quality assurances. The main advantage of the pairing of AI drones and blockchain technology is to provide food products an end-to-end traceability solution where a farm to plate traceability journey is possible. Each user in the food supply chain system from consumers to retailers as well as regulators gain in the real time access of information on the sourcing, processing, and distribution of products reducing food fraud/threat to food safety with contaminated products Blockchain technology brings transparency and efficiency of supply chain by enabling automated smart contracts for the execution of transaction delivery of products, minimizing the procedural paper requirements, and eliminating inefficient intermediary chains. Future studies should focus on scaling and optimizing how AI drone data collection for blockchain networks while addressing issues such as speed of collection, timing of transfers, management of data and security. The combination of AI drones and blockchain technology has the potential to fundamentally change food safety, ethical or sustainable sourcing, and accountability within a global food supply chain, (Patel et al., 2022).

Figure 9. The future of drone technology

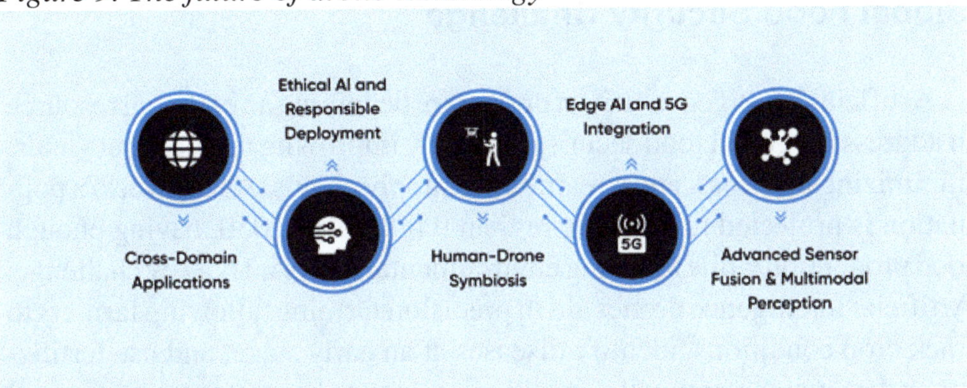

Environmental Impact and Long-Term Sustainability Analysis

Future research on the application of AI drones in the food sector must address long-term environmental sustainability factors. While drones improve efficiency and have potential environmental benefits through precision agriculture, which reduces waste of resources and chemicals and carbon footprint, more research is required to examine the overall environmental impact of drones. The life cycle impacts from drone technology, like energy consumption, battery dismantling, as well as materials use for product durability, ought to be discussed. One area of research involves designing more sustainable drones, which may involve examples like solar-powered drones or model drones using biodegradable materials (Patel et al., 2022) in order to limit potential effects on the environment. Future researchers should look at the contextual long-term effects of drone usage in relation to biodiversity, and soil and ecosystem health. Long-term sustainability studies should include an element around the models and best practices regarding the prudent use of AI-enabled drones in food production. Research on these areas can formalize a means for research to create a framework to maximize the operational benefits of AI drone technology as a sustainable alternative while minimizing potential impacts, to build a safer more resilient sustainable food supply.

The Role of AI Drones in Addressing Global Food Security Challenge

Artificial intelligence-driven drones are becoming an essential resource in addressing global food security issues by improving agricultural yields, minimizing food loss, and streamlining supply chains. As the world's population is projected to reach more than 9 billion by 2050, having enough food production while reducing environmental footprint is a key challenge. Artificial intelligence drones aid in precision farming, allowing farmers to track crop conditions, identify diseases at an early stage, and use fertilizers and pesticides optimally, resulting in greater harvests and minimized wastage of resources, (Chamara et al., 2020). Also, AI drones enhance disaster resiliency through the evaluation of damage caused by droughts, floods, and pests, thereby facilitating quick response and enhanced disaster preparedness, (Patel et al., 2022). In food supply chain logistics, drones optimize last-mile delivery such that food gets delivered to inaccessible and remote regions, minimizing transportation inefficiency-led losses. Future developments in AI-based drone technology, including autonomous swarm drones and real-time analytics integration, will continue to boost food security by increasing agricultural production to be more sustainable, efficient, and climate change and global supply chain disruption resilient.

CONCLUSION

Summary of Key Findings and Insights

This research points to the revolutionary potential of AI-equipped drones in the food sector, specifically in making food production more efficient, sustainable, and secure. AI-operated drones have transformed precision agriculture, maximizing the use of resources by allowing precise irrigation, fertilization, and pest control, thus boosting yields while reducing wastage. Their application in food logistics has enhanced supply chain efficiency by minimizing delays, avoiding food wastage, and providing enhanced traceability using blockchain technology. The study also highlights the economic and environmental advantages of AI drones, where they show a drastic reduction in labour costs, use of resources, and transport costs while reducing carbon footprints and encouraging environmentally friendly agriculture, (Kapoor et al., 2023; Iqab, 2024). Regulatory limitations, initial high prices, and technical issues are, however, major hurdles to widespread usage.

Closing Remarks on the Future of AI-Powered Drones in the Food Industry

Furthermore, AI-enabled drones combined with new-generation technologies such as IoT, Blockchain, or clean technologies will facilitate advancement and sustainability in the rapidly changing realities that influence food systems. As science, research, and scaling become commonplace with the use of once futuristic drones in the agri-food sector, AI-enabled drones will expand its adoption. will affect agri-food production in the short term and are projected to support greater productive, resilient, and sustainable global food system.

REFERENCES

Ahmadi, A. (2023). The AI food revolution: Reshaping food sciences through artificial intelligence. *International Journal of BioLife Sciences*, *2*(1), 62–71. DOI: 10.22034/jbs.2023.178662

Aipoh, G., Osunkanmbi, A. A., & Egbemhenghe, J. (2025). Artificial intelligence in logistics and distribution: The function of AI in dynamic route planning for transportation, including self-driving trucks and drone delivery systems. *World Journal of Advanced Research and Reviews*, *25*(2), 155–167. DOI: 10.30574/wjarr.2025.25.2.0214

Carmona-Torres, C., Dev, K., Mhlanga, D., & Aït-Kaddour, A. (2024). From Food Industry 4.0 to Food Industry 5.0: Identifying technological enablers and potential future applications in the food sector. *Comprehensive Reviews in Food Science and Food Safety*, *23*(6), e370040. DOI: 10.1111/1541-4337.70040 PMID: 39437193

Chamara, R. M. S. R., Senevirathne, N. N., De Silva, S. H. N. P., Ariyaratne, W. M. T. P., & Marambe, B. (2020). Role of artificial intelligence in achieving global food security: A promising technology for the future. *Sri Lanka Journal of Food and Agriculture*, *6*(2), 43–70. DOI: 10.4038/sljfa.v6i2.88

Creswell, J. W., & Poth, C. N. (2018). *Qualitative Inquiry and Research Design: Choosing Among Five Approaches* (4th ed.). SAGE Publications.

EASA. (2024). Easy Access Rules for Unmanned Aircraft Systems. https://www.easa.europa.eu/sites/default/files/dfu/D0593E_2024-07-10_06.26.37_EAR-for-Unmanned-Aircraft-Systems.pdf

Federal Aviation Administration (FAA). (2023). *Part 107 Small Unmanned Aircraft Systems Regulations*. https://www.faa.gov/newsroom/small-unmanned-aircraft-systems-uas-regulations-part-107

Grewal, D., Guha, A., & Noble, S. M.. (2024). The food production–consumption chain: Fighting food insecurity, loss, and waste with technology. *Journal of the Academy of Marketing Science*, *52*, 1412–1430. DOI: 10.1007/s11747-024-01040-x

Grobbelaar, W., Verma, A., & Shukla, V. K. (2021). Analyzing human robotic interaction in the food industry. *Journal of Physics: Conference Series, 1714*(1), 012032. DOI: 10.1088/1742-6596/1714/1/012032

Hwa, L. S., & Chuan, L. T. (2024). A brief review of artificial intelligence robotics in the food industry. *Procedia Computer Science, 232*, 1694–1700. DOI: 10.1016/j.procs.2024.01.167

Iqab, M. (2024). *Harnessing drones for faster, cheaper, and greener logistic solutions in challenging environments* (Bachelor's thesis). LAB University of Applied Sciences. https://urn.fi/URN:NBN:fi:amk-202405039113

Jahani, H., Khosravi, Y., Kargar, B., Ong, K.-L., & Arisian, S. (2024). Exploring the role of drones and UAVs in logistics and supply chain management: A novel text-based literature review. *International Journal of Production Research*. Advance online publication. DOI: 10.1080/00207543.2024.2373425

Kachouei, M. A., Kaushik, A., & Ali, M. A. (2023). Internet of Things-enabled food and plant sensors to empower sustainability. *Advanced Intelligent Systems, 5*(12), 2300321. DOI: 10.1002/aisy.202300321

Kapoor, A., Vo, D.-V. N., & Rangasamy, G. (2023). Digitalization of the agro-food sector for achieving sustainable development goals: A review. *Sustainable Food Technology, 1*, 783–802. DOI: 10.1039/d3fb00124e

Keogh, J. G., Dube, L., Rejeb, A., Hand, K. J., Khan, N., & Dean, K. (2020). The future food chain: digitization as an enabler of Society 5.0. In *Building the Future of Food Safety Technology* (pp. 11–38). Elsevier., DOI: 10.1016/b978-0-12-818956-6.00002-6

Khayyam, M., Ihtisham, M., & Asongu, S. A. (2021). Potential role of technology innovation in transformation of sustainable food systems. *Revista de Agricultura (Piracicaba), 11*(984). Advance online publication. DOI: 10.3390/agriculture11100984

Kitonsa, H., & Kruglikov, S. V. (2018). Significance of drone technology for achievement of the United Nations sustainable development goals. *R-economy, 4*(3), 115–120. DOI: 10.15826/recon.2018.4.3.016

Kitonsa, H., & Kruglikov, S. V. (2018). Significance of drone technology for achievement of the United Nations sustainable development goals. *R-economy*, *4*(3), 115–120. DOI: 10.15826/recon.2018.4.3.016

. Min, W., Zhou,Wen, M., Jiang, S., & Jain, R. (2024). From plate to production: Artificial intelligence in modern consumer-driven food systems. *Journal of Food Systems and AI, 12*(3), 45-62. https://doi.org/DOI: 10.1109/JFSAI.2024.0123456

Miranda, J., Ponce, P., Molina, A., & Wright, P. (2019). Sensing, smart and sustainable technologies for Agri-Food 4.0. *Computers in Industry*, *108*, 21–36. DOI: 10.1016/j.compind.2019.02.002

Patel, M., Sinha, A., & Desai, R. (2022). Weather-based forecasting in agricultural logistics. *Applied Artificial Intelligence*, *36*(1), 45–63.

Suresh Babu, C. V. Koti Reddy, Chandu, G., Kowshik, T., & Rao, L. N. (2024). Trustworthy AI for Optimizing Agriculture: Leveraging IoT Sensory Data. In S. Sharma, A. Prakash, & V. Sugumaran (Eds.), Developments Towards Next Generation Intelligent Systems for Sustainable Development (pp. 124-147). IGI Global. https://doi.org/DOI: 10.4018/979-8-3693-5643-2.ch005

Suresh Babu, C. V., Yadavamuthiah, K., Abirami, S., & Kowsika, S. (2023). Artificial Intelligence in Wastewater Management. In Shikuku, V. (Ed.), *Artificial Intelligence Applications in Water Treatment and Water Resource Management* (pp. 31–45). IGI Global., DOI: 10.4018/978-1-6684-6791-6.ch002

Wang, W., & Siau, K. (2019). Artificial Intelligence, Machine Learning, Automation, Robotics, Future of Work and Future of Humanity: A Review and Research Agenda. *Journal of Database Management*, *30*, 61–79. DOI: 10.4018/JDM.2019010104

Yin, R. K. (2018). *Case Study Research and Applications: Design and Methods* (6th ed.). SAGE Publications.

Zhang, S., Khayyam, M., Ihtisham, M., & Asongu, S. A. (2021). Potential role of technology innovation in transformation of sustainable food systems. *Revista de Agricultura (Piracicaba)*, *11*(984). Advance online publication. DOI: 10.3390/agriculture11100984

Chapter 7
Plant Disease Detection:
A Key to Ensuring Food Security

Kavita Saini
https://orcid.org/0000-0003-4058-8622
Galgotias University, India

Shrikant Patel
Delhi Skill and Entrepreneurship University, India

Palak Saini
Sharda University, India

ABSTRACT

Plant diseases pose a serious risk to the world's food security, resulting in large-scale financial losses and negatively impacting the lives of millions of people globally. Plant disease diagnostics and early identification are essential for efficient oversight and administration. Conventional approaches to plant disease Diagnostic methods including visual inspection and laboratory testing are labor-intensive, time-consuming, and sometimes inaccurate. The evolution of Machine learning-based methods for identifying plant diseases can greatly enhance diagnostic speed and precision, empowering farmers and researchers should act promptly to stop the spread of illness and minimize agricultural losses.

DOI: 10.4018/979-8-3693-9884-5.ch007

1. INTRODUCTION

1.1 Overview of plant diseases and their consequences

Plant diseases are a major threat to global food security, causing significant financial losses and having a detrimental effect on the lives of millions of people worldwide (Mohanty et al., 2016). Plant diseases are thought to be responsible for 20% of crop losses globally each year, according to the Food and Agriculture Organization (FAO). These losses impact the food supply and raise the likelihood of hunger, poverty, and food insecurity. The chapter talks about the early detection of plant diseases and further, chapter is discussing how to control of plant diseases.

Early detection and diagnosis are critical to the management and control of plant diseases. Traditional methods of identifying plant diseases, like visual inspection and laboratory testing, can be time-consuming, labor-intensive, and imprecise. Farmers and researchers will be able to react swiftly and precisely to diagnoses brought about by the development of machine learning-based techniques for plant disease identification, which will help to prevent the spread of diseases and reduce crop losses (Ferentinos, 2018).

The objective of this study is to develop a machine learning-based system for plant disease identification using convolutional neural networks (CNNs). The project's objectives are as follows: Build a CNN model that can accurately identify plant diseases based on visual traits captured in photographs (Brahimi et al., 2017). Examine the model's performance with a large collection of plant disease images.

Examine the model's performance in comparison to traditional methods of disease diagnosis.

Identify potential improvements and future directions for the development of machine learning-based illness detection systems (Food and Agriculture Organization, 2021).

The work presented in the chapter may help to improve global food security and lessen the negative effects of plant diseases on the economy and society by making a substantial contribution to the development of more effective and efficient technologies for plant disease identification.

2. PRESENT TECHNIQUES FOR PLANT DISEASES IDENTIFICATION

Traditional methods have primarily relied on expert visual inspection to identify plant diseases. This approach examines the external characteristics of plants, including color, texture, and the presence of lesions or spots, in order to identify potential diseases. This approach may work, despite being time-consuming, labor-intensive, and often requiring specialized knowledge.

Furthermore, especially when it comes to early-stage ailments or diseases with similar symptoms, eye inspection is not always objective and is prone to errors (Sladojevic et al., 2016). Laboratory-based diagnostic techniques such as polymerase chain reaction (PCR) and enzyme-linked immunosorbent assay (ELISA) have also been used in the identification of plant diseases.

These methods involve collecting plant samples and analyzing them in a lab. These techniques produce extremely accurate data, but because of their high cost, requirement for specialized equipment and trained personnel, and possibility for time-consuming processes, they are not as practical for large-scale disease surveillance (Babatunde, 2024; Hassanien & Darwish, 2020; Kouadio, 2023; Pacal, 2024).

2.1. Prior investigations and studies in this field

In recent years, there has been a surge in interest in using computer vision and machine learning techniques for plant disease identification. The potential of these techniques to improve the accuracy, economy, and efficiency of disease diagnosis has been the subject of numerous studies.

In one of the first works in this field, Mohanty et al. (2016) classified 14 distinct plant diseases from leaf photos using a deep learning system. With a test set, the researchers' Subsequent research has expanded on this work by investigating other machine learning algorithms, picture preparation methods, and dataset sizes. For instance, Ferentinos (2018) achieved a 99.53% accuracy rate in classifying 26 distinct plant diseases using a convolutional neural network (CNN). Brahimi and others al. (2017)

achieved a 99.18% classification accuracy by combining deep learning and transfer learning to diagnose nine distinct tomato illnesses.

2.2. The shortcomings of the present strategies

The demand for better ones Even though plant disease identification using current machine learning-based techniques has shown encouraging results, there are still a number of issues that need to resolve.

Lack of vast, diversified, and well-annotated datasets for machine learning model training is one of the primary problems. A lot of the research that has already been done has employed datasets that are either very small or have little variety, which can cause overfitting and poor generalization to real-world situations. Furthermore, creating high-quality datasets might be challenging due to the time-consuming and error-prone nature of the annotation process for plant disease photos. To address changes in illumination, camera angles, and plant growth phases, better picture preprocessing and data augmentation approaches are required. This presents another issue (Singh et al., 2023). Simple preprocessing methods like picture scaling and normalization have been employed in many previous research, but they might not be enough to manage the complexity of real-world plant disease images.

Finally, to manage the complexity of plant disease detection, better model designs and training approaches are required. Deep learning models can detect plant diseases with high accuracy, according to research that has already been done, but there is still space for development in terms of model complexity, training effectiveness, and interpretability.

Future studies in plant disease identification should concentrate on creating bigger and more varied datasets, enhancing methods for picture preprocessing and data augmentation, and investigating increasingly sophisticated machine learning architectures and training plans in order to overcome these obstacles (Prajapati et al., 2023). More multidisciplinary cooperation between agronomists, computer scientists, and plant pathologists is also required to guarantee the viability, scalability, and efficacy of machine learning-based methods for plant disease identification in actual environments.

3. PREPROCESSING AND DESCRIPTION OF THE DATASET

3.1. Source and amount of data

Plant Village is one of the sources of the plant leaf photos that make up the dataset utilized in this research. With 70,295 photos in the training set and 17,572 pictures in the validation set, the dataset has a total of 87,867 photographs. The photos are divided into 38 groups, which correspond to different plant types and disease states.

3.1.1. Resizing, normalizing, and enhancing images

The photos in the dataset were shrunk to a consistent 128x128 pixel size in order to prepare it for training. In order to feed photos into the CNN model, it is necessary to guarantee that they are of the same size. In addition, each pixel value in the photos was normalized by dividing it by 255, resulting in a range of pixel values between 0 and 1. The training process's stability and convergence are enhanced by this normalization.

The code that was given did not specifically address data augmentation strategies. To lessen overfitting and boost variety in the training set, deep learning practitioners frequently employ data augmentation (Prajapati et al., 2023). One can use methods like flipping, cropping, random rotation, and color jittering.

3.2. Architecture design of CNN

Multiple convolutional layers, max pooling layers, dropout layers, and dense layers make up the CNN model that was employed for this research. The architecture's purpose is to identify pertinent characteristics from the input photos and categorize them according to the right plant disease groups.

3.2.1. ReLU-activated convolutional layers

Initially, a convolutional layer in the model receives input pictures with dimensions of **128x128x3.** This layer employs ReLU activation function and applies 32 3x3 filters with 'same' padding (Ferreira, 2024). The ReLU activation adds non-linearity to the model, while the padding makes sure that the output feature maps have the same spatial dimensions as the input.

3.2.2. Maximum layer pooling

A max pooling layer is utilized to minimize the spatial dimensions of the feature maps and introduce translation invariance after every two convolutional layers. The feature maps are essentially down sampled by a factor of 2 in both the height and width dimensions when using the max pooling layers, which employ a pool size of 2x2 and a stride of 2.As the model gets deeper, the max pooling layers assist to capture more abstract and high-level characteristics while simultaneously lowering computational cost and the chance of overfitting (Alhammad, 2025).

The Flatten layer is used to flatten the feature maps into a 1D vector following the convolutional and max pooling layers. After that, a thick layer with 1500 units and ReLU activation passes across this vector. The ReLU activation adds non-linearity and facilitates the learning of intricate decision limits by the model.

The model's output layer, which contains 38 units in total, represents the number of classifications of plant diseases (Alhammad, 2025). Using the softmax activation function, this layer creates a probability distribution across the classes using the output logits. For the input picture, the class with the highest probability is used as the predicted label.

3.3.1. Adam optimizer with 0.0001 learning rate

The Adam optimizer with a learning rate of 0.0001 is used to construct the model. Adam is a well-liked optimization technique that works well for deep neural network training because it combines the advantages of momentum and adjustable learning rates. Because the learning rate of

0.0001 is less than the default value of 0.001, the training process is stabilized and significant adjustments to the model parameters are avoided (Sambana, 2025). Although it may take longer to converge, a model with a lower learning rate may be more accurate and reliable.

3.3.2. The loss of categorical cross-entropy

The model was trained using categorical cross-entropy as the loss function. This loss function, which calculates the difference between the actual one-hot encoded labels and the expected probability distribution, is frequently applied to multi-class classification problems.

By reducing the categorical cross-entropy loss, the model is more likely to correctly categorize the plant disease photos, giving high probability to the right classes and low probabilities to the wrong ones.

3.3.3. Ten epochs of training

The training set is used to train the model for ten epochs. An epoch is characterized as a whole run through the training set. The model modifies its parameters at each epoch using the gradients that were calculated from the training set of data.

The purpose of the validation set is to keep an eye on the model's performance throughout training and guard against overfitting (Sambana, 2025). The model's performance is assessed on the validation set after every epoch, and training is terminated when either the target performance is attained or the validation loss stops getting better.

3.4. Metrics for performance evaluation

Classification accuracy serves as the project's main performance indicator. While the validation accuracy is determined by utilizing the validation set, the training accuracy is determined by assessing the model on the training set.

The model can accurately categorize the training samples, as seen by the training accuracy of 99.38%. The validation accuracy is 97.01%, indicating that the model can correctly categorize photos of plant diseases and generalizes well to new data. The following other assessment measures are possible for this project: area under the receiver operating characteristic (ROC) curve, F1-score, precision, and recall. These metrics offer a more thorough evaluation of the model's performance, particularly in the case of unbalanced datasets or particular interest classes.

Confusion Matrix

A confusion matrix for plant disease prediction shows how well a classification model distinguishes between healthy plants and different disease categories. It's a square matrix comparing actual labels vs. predicted labels.

Figure 1. Plant Disease Prediction Confusion Matrix

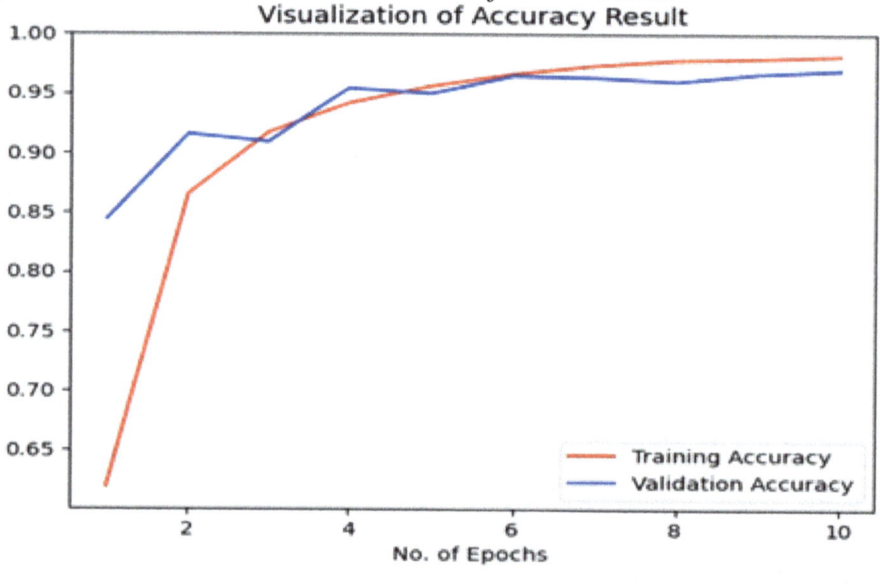

4. SPECIFICS OF IMPLEMENTATION

The following phases comprise the installation of the CNN-based plant disease detection system:

4.1. Preprocessing of Data

With the aid of tf.keras.utils.image_dataset_from_directory, the training and validation datasets were loaded. The photos were scaled to 128 by 128 pixels, batch-sized at 32, then normalized by dividing each pixel value by 255. There were 70,295 photos in 38 classes in the training set and 17,572 images in the validation set.

4.1.2. Architecture of the Model

CNN was constructed with tf.keras.models.It is sequential. It was composed of max pooling layers after many convolutional layers with ReLU activation:

One max pooling layer, two convolution layers with 32 filters, and a 3x3 kernel One max pooling layer, two convolution layers with 64 filters, and a 3x3 kernel One max pooling layer, two convolution layers with 128 filters, and a 3x3 kernel One max pooling layer, threex3 kernel, and two convolution layers with 256 filters One max pooling layer, threex3 kernel, and two convolution layers with 512 filters

Layers of dropouts were used to lessen overfitting. After the output was flattened, a thick layer containing 1500 units and ReLU activation was applied. There were 38 units in the final output layer with softmax activation, which matched the number of classes.

4.1.3. Validation and Training

With a learning rate of 0.0001, the Adam optimizer, and categorical cross-entropy loss, the model was assembled. The validation set was utilized for assessment after ten epochs of training on the training set. 99.38% was the training accuracy and 97.01% was the validation accuracy.

4.2. Outcomes

When the trained model was assessed using the training set, it had a 99.38% accuracy rate. With a validation set accuracy of 97.01%, the model is capable of reliably classifying plant disease pictures and demonstrating good generalization to new data.The great accuracy of both training and validation show how well the CNN-based method for plant disease detection works. With a high degree of accuracy, the model was able to identify pertinent aspects from the photographs and categorize the pictures into the appropriate illness categories.The system that has been created has the potential to be a useful tool for the early diagnosis and detection of plant diseases, allowing farmers and researchers to take prompt action to stop the spread of disease and lower crop losses, based on the implementation details and results.

4.2.2.Snapshots of Interfaces

Figure 2. Interface

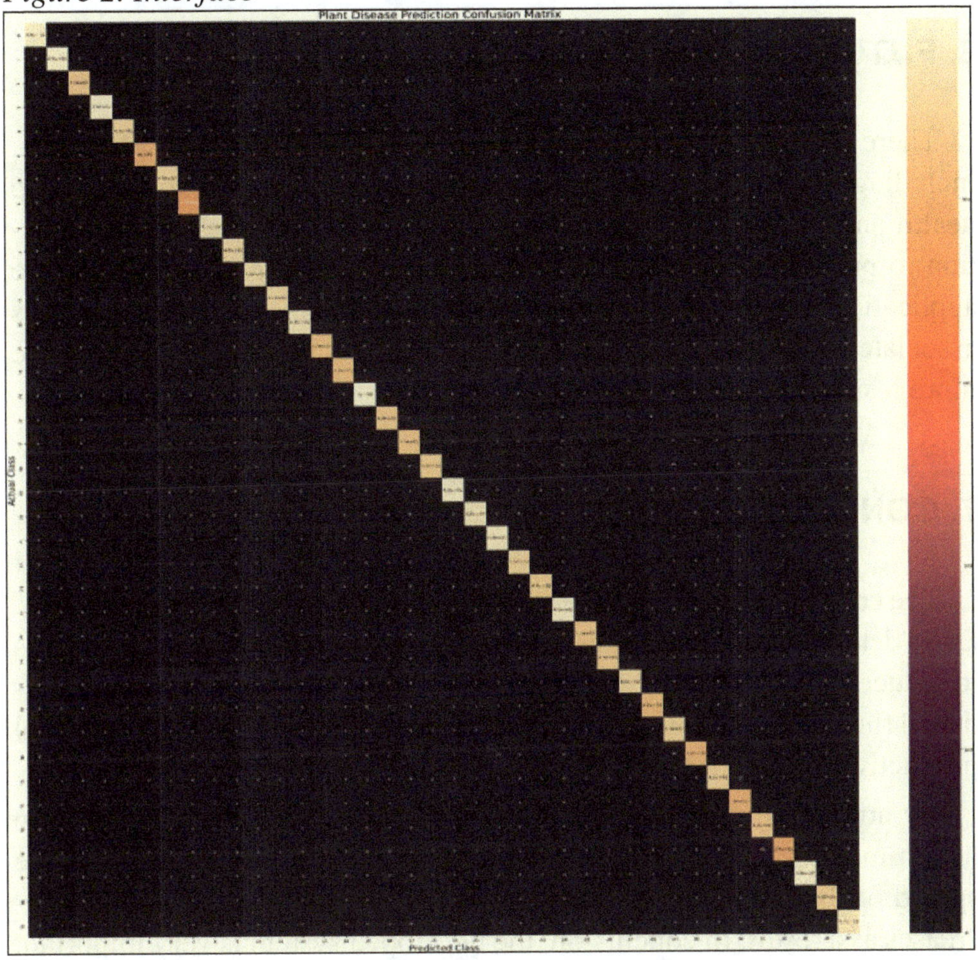

5. RESULT

The research's objective was to develop a machine learning-based plant disease detection system using convolutional neural networks (CNNs). After being trained on a large dataset of plant leaf pictures, the system achieved a high accuracy of 99.38% on the training set and 97.01% on the validation set. The presented work successfully classified pillnesses based on the visual features extracted from the images.

6. FUTURE SCOPE

There are several agricultural applications for the developed system, including precision agriculture by employing technology to track crop health and identify diseases in real-time, farmers may take focused action to prevent the spread of diseases. Crop breeding is another future application. Where the presented work may be enhanced to identify genes associated with disease resistance, breeders can produce crops that are more resilient to diseases.

7. CONCLUSION

The current system will have a major effect on the agriculture industry. Early detection of plant diseases allows farmers to take timely measures to reduce crop losses and stop the disease's spread. This could lead to reduced financial losses, improved food security, and increased agricultural productivity.

By utilizing the presented work real-time disease detection and crop health monitoring features, farmers can respond quickly to prevent the spread of diseases and reduce crop losses. Food security eventually increases as a result. In conclusion, the developed system is very important for the agriculture industry and can improve food security, reduce crop losses, and boost crop yields.

REFERENCES

Alhammad, S. M. (2025). *Deep learning and explainable AI for classification of plant diseases*. Front. AI.

Babatunde, R. S. (2024). *A novel smartphone application for early detection of plant diseases. (app/prototype & field study)*. Academic Press.

Brahimi, M., Boukhalfa, K., & Moussaoui, A. (2017). Deep learning for tomato diseases: Classification and symptoms visualization. *Applied Artificial Intelligence*, *31*(4), 299–315.

Ferentinos, K. P. (2018). Deep learning models for plant disease detection and diagnosis. *Computers and Electronics in Agriculture*, *145*, 311–318.

Ferreira, L. C. (2024). Hyperspectral imaging for detection of plant pathogens — Review and advances. *Front. Plant Sci.*

Food and Agriculture Organization. (2021). The State of Food Security and Nutrition in the World. FAO.

Hassanien, A. E., & Darwish, A. (Eds.). (2020). *Machine Learning and Artificial Intelligence in Agricultural Research*. Springer.

Kouadio, L. (2023). A review on UAV-based applications for plant disease detection. *Remote Sensing*.

Mohanty, S. P., Hughes, D. P., & Salathé, M. (2016). Using deep learning for image-based plant disease detection. *Frontiers in Plant Science*, *7*, 1419. PMID: 27713752

Pacal, I. (2024). *A systematic review of deep learning techniques for plant disease detection (2020–2024)*. Academic Press.

Prajapati, J. B., Barad, R., Patel, M. B., Saini, K., Prajapati, D., & Engineer, P. (2023). Smart farming ingredients: IoT sensors, software, connectivity, data analytics, robots, drones, GIS-GPS. In *Applying drone technologies and robotics for agricultural sustainability* (pp. 31–49). IGI Global.

Sambana, B. (2025). An efficient plant disease detection using transfer learning (YOLOv7/8). Sci. Rep. / Nature Communications type article.

Singh, J. N., Saxena, S., Saini, R., Chauan, M., & Saini, K. (2023, December). The Role of Artificial Intelligence in Information Acquisition. In *2023 5th International Conference on Advances in Computing, Communication Control and Networking (ICAC3N)* (pp. 1153-1157). IEEE.

Sladojevic, S., Arsenovic, M., Anderla, A., Culibrk, D., & Stefanovic, D. (2016). Deep neural networks based recognition of plant diseases by leaf image classification. *Computational Intelligence and Neuroscience*. PMID: 27418923

Chapter 8
Smart Dining:
AI-Powered Innovations Shaping Customer Experience in the Food Industry

R. Siva Subramanian
https://orcid.org/0000-0002-7509-9223
SRM Institute of Science and Technology, India

D. Jayasutha
Kathir College of Engineering, India

Vivek Pandiya Raj
https://orcid.org/0000-0001-8036-8603
Vel Tech Rangarajan Dr. Sagunthala R&D Institute of Science and Technology, India

G. Porkodi
Vel Tech Multi Tech Dr. Rangarajan Dr. Sankuthala Engineering College, India

M. Gowri
Sathyabama Institute of Science and Technology, India

T. P. Anish
https://orcid.org/0000-0002-4308-9645
R.M.K. College of Engineering and Technology, India

M. Nalini
https://orcid.org/0000-0001-8987-3120
S.A. Engineering College, India

ABSTRACT

The integration of Artificial Intelligence (AI) in the food industry is revolutionizing customer experience by enabling personalized services, real-time interactions, and operational efficiency. This chapter explores how AI-driven technologies such as machine learning, natural language processing, computer vision, and recommender systems are enhancing the customer journey across food delivery, restaurant services, and backend operations. The chapter explores applications like chatbots, voice assistants, visual recognition, and sentiment analysis, while addressing key

DOI: 10.4018/979-8-3693-9884-5.ch008

challenges including data privacy, algorithmic bias, and regulatory compliance. The chapter also covers the topic of ethical issues and sustainability and such topics as generative AI, emotion recognition, and federated learning. Thus, this chapter aims to give a brief on how AI can be employed in the food industry to develop smart solutions that enhance customers' experiences in the changing industry environment.

INTRODUCTION

Background on Customer Experience in the Food Industry

Customer experience (CX) has been an essential part of the food industry, which refers to all the processes involving a client that begins with menu perusal and ordering to receiving food and giving feedback. In restaurants, fast foods, cafes, and online food delivery services, it is crucial to ensure that the customers get the best experience to ensure they stick to your business and may recommend others to do the same (Sudiyono et al 2022). As the competition in the food sector is quite high, even small failures in service, delay in delivery or absence of individual attention to the customer leads to loss of the customer. In the past, customer experience in the food industry was handled by people and their feeling or instinct. Managers and owners of restaurants and restaurants depended on face-to-face communication, questionnaires, and their observations. But as the customers get wiser, the services get more complex, and there is the emergence of digital technologies, the traditional ways prove to be inadequate. The modern consumer is an informed, connected, and connected one who seeks personalized, quick, and homogeneous services across channels. To meet these requirements, the food industry has slowly entered the process of digitalization. Mobile applications, point of sale systems, online reviews, and delivery aggregators have been already the driving force of change. But it is the idea of creating a novel form of customer experience with the help of AI the science that can bring change to the businesses and become the key to winning in a digital-savvy environment.

Rise of AI Technologies Across Industries

AI is not something new or exclusive to large enterprises and research institutions; today, it is an established technology implemented in healthcare, finance, retail, transport and educational sectors. As we have seen, with the help of machine learning, deep learning, computer vision, and NLP AI systems are able to do the things that previously only a human mind could do, including understanding language, recognizing images, predicting behavior and making decisions. Today, the use of AI is widespread in the food industry in particular (Yamini et al 2023). It is applied to the processing of big customer data, automated orders through natural language processing, forecasting demand, supply chain management and even the cooking process through smart utensils (Hwa et al 2024). This is making it possible for food businesses to provide better, faster, and more entertaining services to their customers. For instance, it can use a customer's previous orders, his/her preferred diet, and his/her social media profile to suggest meals. The customer can be reached out at any one time through the chatbots while Predictive analysis makes it possible to plan and ensure that delivery of the products or services is done within the shortest time possible. Currently, even physical restaurants HBs are changing customer experience through AI aided Kiosks, smart kitchen, voice interface, and many more. Based on the characteristic features of the food industry as a dynamic setting that depends on a number of preferences and depends on several factors, AI can give an immediate response, learn new data and patterns, so it can effectively contribute to improving the image of the industry for the customer.

Motivation and Scope of the Chapter

The rationale for this survey chapter is to discuss the current state of the use of AI technologies in enhancing customer experience in the food industry. Though there are many papers on AI in general or AI in retail, there is a lack of papers that address the role of AI in micro level as a means to enhance the customer experience in the context of food services. The food industry is a very broad area in which many challenges and opportunities for the application of artificial intelligence can be found. It is a combination of the online and offline approaches as well as a broad spectrum of customer demands that have to be addressed, immediate customer workflow and a focus on customization and quality. To better comprehend how these aspects are being tackled with the use of AI technologies the virtual assistants, recommendation systems, emotion-aware chatbots, or intelligent kitchens, for example it is essential to conduct a sectoral approach. Specifically, this chapter seeks to fill that gap by integrating current information, technology, and trends on AI in enhancing customer experience in the food industry. It analyses how the primary application fields of artificial intelligence can improve customer experience and loyalty, ones to be addressed are chat bots and voice interfaces, personalized suggestions and real-time delivery management. It also looks into how such back-end applications affecting its inventory, quality control, and other related sectors, are indirectly enhancing customer experiences. In addition, this chapter aims at discussing the limitations and drawbacks of applying AI in the food industry and some of the problems associated with data privacy, model bias, cost of integration and customer trust. Indeed, it also underscores the necessity of compliance and ethical regulation of AI as well as achieving sustainable development. Last but not the least, the chapter enlists the future research areas and trends including generative AI, multimodality, and AI with Blockchain and IoT for enhanced security and transparency along with adaptability in the customer experience.

Contributions and Structure of the Chapter

The following are the four contributions of this survey chapter:

Literature Review

It provides a detailed discussion of the current state of AI technologies that are employed to improve customer experience in the different areas of the food sector such as online food delivery services, restaurants, cafes, and cloud kitchen.

Taxonomy of AI Applications

The chapter provides a classification of AI applications that are divided into functional areas like customer interaction, personalization, sentiment analysis, back-end optimization, etc., which would help readers understand the role of AI in various fields.

Challenges

It presents the primary difficulties that food businesses encounter when adopting AI-driven customer experiences, technical, ethical, and operational, as well as possible solutions and recommendations.

Future-Oriented Perspective

Thus, the chapter offers a forward-looking perspective by pointing out future research directions and trends that would help scholars, policymakers, and practitioners in their work.

The Rest of this Chapter is as Follows

Section 2 gives an essential background information on the main AI technologies applied in the customer experience applications such as machine learning, natural language processing, computer vision, and IoT. Section 3 goes further to describe the major AI applications that directly impact customers in the organization by including recommendation systems, AI chatbots, delivery forecasts and sentiment analysis. Section 4 focuses on the support back-end technologies of supply chain, demand forecasting, and smart kitchen technologies that indirectly impact the customer. In section 5 of the article, the author discusses the risks and drawbacks that AI implementation in the food industry has: data security issues, ethical questions, and technology constraints. Section 6 covers regulatory frameworks, ethical considerations, and the growing focus on sustainable AI practices within the industry. Section 7 presents some of the ongoing and future research directions which are the generators AI, Emotion AI and Decentralized data analysis. Section 8 presents the chapter's conclusion, which discusses the study's outcomes and future trends of using AI in customer experience in the food industry.

FOUNDATIONS OF AI IN CUSTOMER EXPERIENCE

AI is now a new way through which organizations communicate and provide services to consumers in various sectors, including the food industry. From forecasting to natural language processing and conversational AI, as well as ROBO, it has become the foundation of most CX strategies (Nguyen et al 2022). To better understand the uses of such technologies in the food industry, it is necessary to define the four core technologies behind them: Machine learning, Natural Language Processing, Computer Vision, and Recommender systems. Besides that, combining with data analytics, IoT and human- AI interaction models makes it more effective and applicable in realistic environments (Pajila et al 2023; Nithya et al., 2023).

Overview of AI Techniques

Machine Learning (ML) is the primary component of AI that can be applied in customer experience. It is important to note that ML utilizes systems that can learn from the data provided without the need for being programmed to do so, (Sreelakshmi et al 2024). In the food industry, using ML, one can predict the customers' preferences, predict the demand, identify some problems, for instance, the delay of the orders, and improve several services (Mahesh, B. 2020). For example, ML models can help in recommending the dishes based on the purchasing history of customers, in dynamic pricing of the food items, or better inventory management. Supervised learning is used in problems like customer churn prediction or categorizing feedback while unsupervised learning is used in grouping customers based on their behavior (Saravanan et al 2023).

Figure 1. AI techniques in the food industry

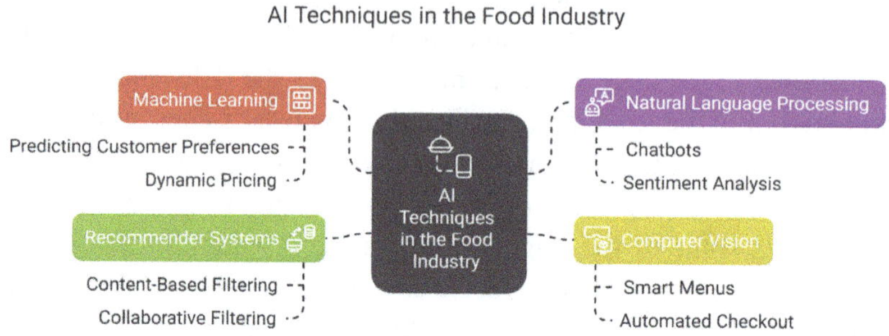

Natural Language Processing (NLP) enables the AI systems to analyze, comprehend, and even generate human's language. In the food industry, NLP is used in the conversational agents such as chat bots and voice agents for ordering food, asking about menu options, or even giving feedback. NLP is also used in the sentiment analysis where people's view on a certain brand or product is determined through analysis of their reviews or posts on the social media. Such features as the named entity recognition, scoring of sentiments, and identification of intents assist food businesses in determining what to do with the received inputs and feedbacks.

Other component is the Computer Vision (CV), which is relevant to improve the visual experience of the customer journey. CV helps to interpret images and video by providing the machine with the ability to understand the contents of images. In the field of food services, it can be applied to smart menus, where the system can identify the food items that have been ordered, estimation of the calorie content of the meals, monitoring of the kitchen and even the emotions of the customers using face detection. CV also includes the automated checkout systems, visual quality check and custom menu on digital kiosks.

Thus, Recommender Systems employ ML, NLP, and collaborative filtering to impact customers by choosing the most suitable items to offer. These recommend items through previous purchasing trends of the users, similar users' choices or even the time of day and geographical location. Recommender systems are widely used in the food delivery applications, restaurant mobile applications, and subscription-based diet services providing the recommendations on the meals, diets, and special offers. Content-based and collaborative filtering techniques are normally combined to create hybrid recommender systems for better results.

Role of Data Analytics and IoT in Enhancing AI Applications

This means that the effectiveness of an AI system depends on the data fed into it for training purposes. This is where data analytics, in conjunction with the IoT-Internet of things comes in handy of providing complementary structure and formats for the data needed for informed decisions.

Data analytics covers the capturing, storing, and analysis of large volumes of data irrespective of the structure type. In the food industry, the client touchpoints are numerous; customers may engage with the food business through applications, websites, social media, physical outlets, and third-party services. Information technology tools consolidate and analyze this information to produce information that can be utilized for decision making. These insights help the AI systems in the learning of patterns and the optimization of the flow and offering of services. For instance, it is possible to examine the peak order time, the most viewed dishes, or the most common complaints to make decisions that improve the satisfaction of the customers.

Analytics is also used in another way in the customer journey mapping, monitoring of KPIs and discovering pain points. The actual and anticipated performance data on the dashboards allows managers to make timely changes to operations, promotions, or personnel to better meet customer traffic.

IoT devices are input and output components of AI systems. Some IoT devices in food service are smart fridges, ovens and cameras, smart clothing, sensors in delivery boxes, and self-service kiosks. These devices capture current data on temperatures, time taken in cooking, stock, delivery time, etc. For instance, smart kitchen utensils may interact with the application to notify workers that such foods are ready for sale because they are well cooked, or notify the management that such food items are scarce.

IoT also enhances customer-facing innovations. For instance, some of the tables in smart restaurants can have touch screens and sensors that can sense when a customer is willing to order food or ask for the waiter. In delivery logistics, IoT through GPS and sensors to track location and even temperature of the food to be delivered, which directly affects customer satisfaction. By connecting IoT with AI, the food businesses can make use of the real time, context-aware recommendation system that can address the needs of a particular customer at the right time and in the right place.

Human-AI Interaction Models in Customer Service

As AI continues to enter the picture and become an influencing factor in customer experiences, it is important to develop a good human-AI interaction model. These models describe how customer interact with AI and how the latter interacts with customers as being empathetic, clear, and useful.

Conversational AI is one of the fields in customer service that uses NLP, ML, and in some cases, speech recognition to enable a conversation with customers. In the food services, there is an engagement of customers with chatbots or voice assistants to book tables, order meals or even lay complaints. These systems need to be designed to handle multi-turn dialogues, identify context, and respond in an appropriate manner for the customers. For example, a chatbot that does not ask the right questions on a customer's allergy or provides a general answer may cause frustration or even endanger the customer's life.

To enhance trust and ease of use, the AI systems should be identifiable as non-human, aware of the context and should have the capability to pass on the queries which are beyond its capability to human agents. One of the key factors that contribute to the successful implementation of customer service automation is the ability to transfer a conversation between the AI and human service staff.

More complex and natural interfaces that include from text, voice, gesture and visual interfaces are being explored in the increase of the interaction models. For instance, a customer can order for a product through a voice command to a smart speaker while there is a digital menu displayed on a screen. AI systems that are capable of processing information in different forms and integrating the information is more natural.

Emotion-aware AI is relatively new field of study where customers' mood or sentiment is determined through their tone of voice, facial expressions or the words they choose to use. It is this emotional intelligence that enables AI to adapt its responses, for instance, to give an apology and some form of compensation when the customer is annoyed or, on the other hand, to try to sell more when the customer is happy. These types of systems greatly improve the perceived emotions of empathy of the AI-based service.

Personalization and adaptive systems are also considered to be the part of human-AI interaction. This is the case of systems that help the user remember his or her preferences or adapt recommendations based on the user's behavior or even the way the system communicates with the user. Such models also help in creating engagement and eventually customer loyalty in the long run.

The primary technologies of AI that are deemed to drive the intelligent customer experience solutions in the food industry include; machine learning, natural language processing, computer vision and recommender systems. These technologies are helpful when backed up by well-developed data analytics and IoT framework, which provide real time, customer specific and context based services that help customers relate differently with their food brands.

Moreover, there are characteristics of human-AI interaction models such as conversational agents, emotion-aware systems, or multimodal interfaces that are important to create positive experiences and meet people's expectations. The advances in AI will always be built upon these elements and will thus be more adaptive and focused on the human in the food industry's customer journey.

AI APPLICATIONS ENHANCING CUSTOMER EXPERIENCE IN THE FOOD INDUSTRY

AI is now a disruptive technology that is revolutionizing how customers in the food industry are being managed. used in recommending system, voice ordering system and many more, artificial intelligence is defining a generation change that is revolutionising the consumers experience in identification process of foods services. This section covers the main AI applications that improve customer experience and categorizes them into six groups.

Figure 2. AI transformations in food customer experience

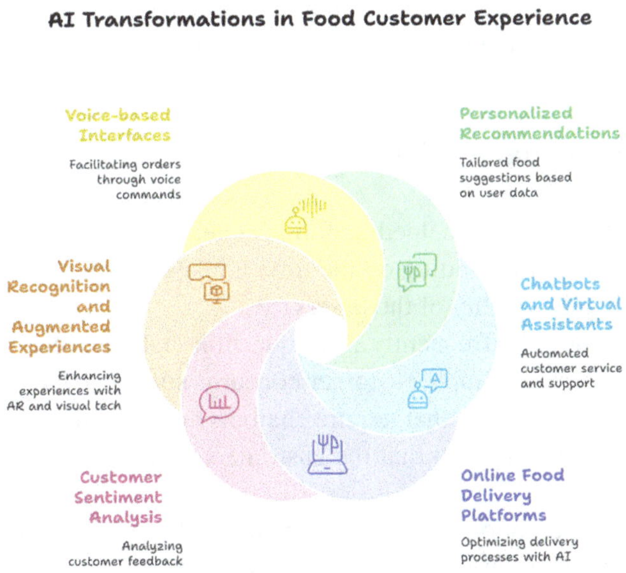

Personalized Recommendations

The use of personalized recommendations has become a norm in food service platforms in the current society. AI systems use information about customers' behavior, their preferences, previous orders, as well as time, place, and even weather to offer the most intelligent food recommendations.

Recommended items in the menu are generated by collaborative and content-based filtering which aims at presenting foods that the user may prefer. This can lead to increased satisfaction and thus more business from the same clients. For example, if a user often prefers vegetarian meals, the AI will recommend similar dishes, but occasionally, it will present new dishes to the user so that the user will not get bored.

Taste and dietary preference modeling goes further by creating an individual user model based on the ingredients, kinds of Flavors, nutritional requirements, and health restrictions. This is because recommendations can be made to change based on real-time data on such parameters as caloric intake or activity levels from fitness apps and wearables.

Customer segmentation using ML enables the grouping of users into the categories based on their behavior. This leads to a custom approach to targeting certain demographics, e.g., promoting healthy meals to user that are associated with fitness or introducing new dishes to the users that are open to trying new dishes. These enhancements also tend to help in the area of customer interaction and retention in a given channel.

Chatbots and Virtual Assistants

Conversational AI has also helped customers interact with the food businesses in a more efficient manner by cutting on the time it takes to receive support services and also improving the quality of those services.

Ordering, responding to frequently asked questions, taking orders for a restaurant or a hotel, and addressing simple customer concerns such as wrong delivery or refund are some of the functions that require chatbots and virtual personal assistants. These systems are always on so that the customers' complaints can be dealt with whenever they arise.

Multilingual and emotion-aware bots are highly beneficial to the community because they are more accessible and empathetic. AI chatbots are also available in multiple languages and dialects so that it can accommodate different customers. NLP enables the bot to detect frustration or satisfaction with the words used by the user, and use the results to either transfer to a human or offer a refund.

Some examples of the applications of conversational AI in restaurants and delivery services are self-checkout kiosks, conversational interface kiosks, virtual assistants within applications, and voice assistants. This helps reduce the use of human employees and yet the turnover rate of clients is enhanced during busy hours (Maheswari, B., & Nagarajan, R. 2024).

AI in Online Food Delivery Platforms

Given the fact that food delivery is an online service, the use of AI is evident and crucial in making the delivery to be efficient, fast, and seamless to the users. It improves not only the backend of the system but also the frontend part which is the user interface.

Order management and the time required to make the delivery are based on machine learning to review historical data, the current load in the kitchen, traffic, and the availability of the courier. These predictions help in controlling the expectations of the customers and in turn, the satisfaction of the customers can be enhanced because of the reduction in delays and uncertainties (Gracious et al 2023).

Suboptimizations for routes and fleets involve the utilization of AI techniques such as Genetic algorithms and Reinforcement learning in order for it to come up with efficient routes and the appropriate delivery people. This not only shortens the delivery time but also leads to the improvement of overall operational costs and efficient use of delivery facilities.

Order status notification services ensure that the customer is updated at every stage of food preparation, courier collection and delivery. AI makes certain that such updates are made in good time and with the right information, which enhances the credibility of the service.

Customer Sentiment Analysis

Perception of customers is critical in ensuring that brand image is well maintained and any service failure is corrected. AI facilitates this through advanced natural language processing (NLP) techniques.

Review and feedback analysis forms part of NLP that helps in analyzing customer reviews, survey responses, and chat interactions to get opinions, trends, and issues. The scores can be employed to determine the level of satisfaction of the customers and make changes to the menu, the quality of services, or the training of the employees.

Another application is in the monitoring of social media and in the evaluation of brand image. Such tools crawl information from social media such as Twitter, Instagram, Facebook, and determine the popularity of specific trends or complaints. This assists the brands in managing their reputation and addressing any cases that may occur before they get out of hand.

Sentiment analysis is also used by automated response systems to reply to the reviews and feedback given by the customers. Praise and recommendations may earn the customer thank you messages and follow-ups, while complaints may get an apology, refund or being passed to a human interface, to make the customer feel valued.

Visual Recognition and Augmented Experiences

In different food business venues, the incorporation of the computer vision and AR technologies has made it possible to provide customers with enhanced, virtual, and visually appealing experience.

Visual menus and food recognition let the clients look at photos of dishes and sometimes add nutritional values or popularity indicators on top of them. Some of the platforms apply AI to analyze and recognize food items from images or videos and recommend or offer substitutes.

Digital menus or experiences through AR/VR help in increasing the in-app usage as the users get the feel of the food directly on their plate through their smart devices. This also assists in decision-making and makes the customers more likely to order items they are not familiar with, thus minimizing the concept of menu anxiety and enhance the chances of upselling.

Voice and gesture based self-ordering surfaces incorporate AI vision to identify the returning guests, and eye tracking or gestures for their choice along with contactless UIs that became popular due to COVID-19. Such systems include increased efficiency, which is also connected to the novelty of the systems, thus improving both the functionality of the systems, and the image of the brands.

Voice-Based Interfaces

With the help of voice interfaces, food businesses are embracing voice interfaces to optimize the communication with customers. Interactive voice ordering and providing feedback is on the rise specifically regarding voice activated smart speakers and car systems. The application enables the customers to order meals, repeat previous orders, or give instant feedback through voice commands. This makes it convenient for users who can interact with the system with their hands busy and also it is convenient for the disabled. You will find more and more drive-thru systems are tend to adopt voice ordering interfaces so as to provide faster, more personalized and less error call ordering. Such systems, integrated with AI, can verify the orders, suggest other products, identify the clients by voice or account number. AI has found its way into the food industry and is changing the way different businesses engage with and meet their customers. Not only these technologies help to improve the delivery of services but it also augments the level of satisfaction and engagement that customers have with the restaurant. The integration of the machine learning, natural language processing, computer vision and voice technologies have offered the customer experience that many of them find much faster, smarter and more intuitive. Nevertheless, the effective implementation of these systems requires data protection, openness, and accountability for artificial intelligence, which will be discussed in the further parts of this survey (Prabha et al 2024).

AI IN BACK-END OPERATIONS IMPACTING CUSTOMER EXPERIENCE

Despite this, most of the cases that are well known and perceived by the consumer offer an AI experience in the food industry are applied behind the scenes, such as chatbots, recommendation systems, and smart menus. It is also changing back-end processes like inventory management, supply chain, demand planning, quality check, and kitchen processes. All the manoeuvres taking place in the background impact the customer in terms of service delivery, food quality, price and health wise when consuming products. This section details four areas that show how AI in back-end operations is vital for the improvement of customer experience in the food industry.

Figure 3. AI in back-end operations impacting customer experience

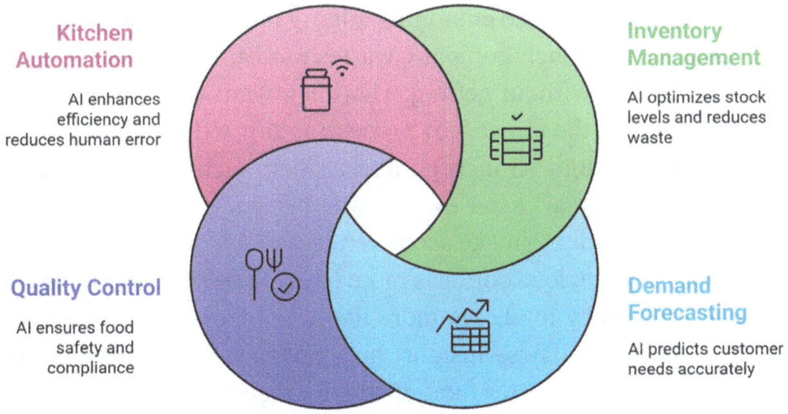

Inventory and Supply Chain Management

Perhaps one of the most significant challenges in managing a business that deals with food is inventory and supply chain management. Lack of smooth flow in these areas results in stock-out, wastage of food items, order delay and unsatisfied customers. AI brings the elements of intelligence and flexibility into this environment. Inventory as a category of AI applications leverages real-time data from POS, IoT-embedded inventory storage, as well as supplier data to track the inventory with the utmost accuracy. It is able to give an indication when some ingredients or products are likely to be exhausted and automatically order them, thus averting a situation where they run out of stock and affect service delivery. It can also categorize the inventory by its perishability, usage frequency, and demand so that the highly perishable products are used first. Whereas computer vision and sensors in this regard improve the capability by detecting spoilage or variation of storage conditions, this helps reduce health risks and wastage. Concerning, the supply chain, AI models analyze supplier reliability, transportation route, as well as the previous delivery data to determine the best vendors and delivery schedules. This helps in replenishing the stock frequently, avoid congestion and enhance the consistency of the product which are essential ingredients in the delivery of quality customer service.

Demand Forecasting and Dynamic Pricing

Customer demand is one of the key aspects that need to be understood and predicted in the food industry. They also entail misjudging the demand which may compel producers to produce too much-company wastes foodstuff or too little- consumers will have to wait longer without getting what they want (Anusha et al 2025). AI improves the accuracy of the forecast by processing big and multidimensional data. Demand forecasting through AI incorporates aspects such as past sales, customer preferences, time of the year, local events, weather conditions, and social media trends among others in order to make accurate predictions of customer demand. For instance, it can classify that iced coffee is more likely to be bought over the weekend in warm months, and spicy foods are more likely to be ordered in cold months in some geographic location. These insights help businesses in staffing, purchasing inventory, and preparing meals for their clients. Forecasting helps one to minimize wastage of food, optimize the usage of resources and enhance the level of service delivery and quality to the customer. Furthermore, it also helps in dynamic pricing that means the prices of the products can be changed according to demand, time, stock, and competitor actions. For instance, during the busy hours or when a particular food item or meal is most ordered, the prices could be slightly raised, while the discounts for a restaurant can be offered during the slow hours. Such changes in the pricing when done in a right and proper manner can be mutually beneficial: businesses get their profit and consumers get more options.

Quality Control and Food Safety Monitoring

These fundamentals are essential for the customer who wants to be assured of getting a quality food product. Any mistake can result in a bad word of mouth, loss of customers, or even legal action being taken against the company. The AI technologies can provide optimal solutions to control the quality of the foods and monitor the food safety. Automated vision systems installed in the kitchen or manufacturing line can be used to visually check the quality, appearance, and uniformity of the food products. Such systems are capable of identifying issues such as overcooking, undercooking, as well as improper presentation of food. For fast food and quick-service restaurants (QSRs), this implies that products produced within the restaurants meet the required standards every time. Temperature sensors and IoT devices are also used in storage conditions for managing temperatures and humidity in fridges and freezers. In case these parameters are out of range, one can set alert to go to the staff members and help avoid spoilage or contamination. Other categories of business intelligence can also track the past performance of equipment and predict future failure. For instance, a fault in an oven or fryer can be easily noted if there are variations in energy consumption or temperature. Minimizing equipment downtime is important to avoid disruption of kitchen operations and to meet customers' orders on time.

Kitchen Automation and Robotics

There is a growing usage of facial recognition technology in food preparation systems and or consumer handling and interaction systems inclusive of system Robotics and Automations techniques used in QSRs, large chains as well as the cloud kitchen settings. All these technologies enhance speed, quality, and cleanliness, which are all factors that will in one way or another affect the customer. Some of the kitchen operations that can be automated using the AI include dough making, chopping of vegetables, portioning and even cooking. For instance, robotic arms can be used in flipping burgers, frying chicken, or even preparing pizzas with the same quality and standard. It can also order food ingredients in a way that they will be ready at the right time for cooking and avoid confusion and time wastage. This kind of automation assists in satisfying the demand at some certain time or period of time. In venues where customers are constantly flowing through the restaurants such as stadiums, food courts or airports, robots are capable of handling hundreds of orders at a time, thereby controlling both throughput and cost of labor. Automated machines such as robots that incorporate visual and touch features is able to monitor the cooking process, look for signs of burn or raw foods and make adjustments in the process. With the help of Artificial Intelligence scheduling systems, they are able to sort out orders, so that timely service can be provided and waiting time can be minimum during peak hours. From the hygiene and safety point of view, they lessen human interaction with the food products, hence, reducing risks of contamination. This is especially so in the current world and especially after the COVID-19 pandemic outbreak where issues to do with contactless food preparation and clear measures that show that food is prepared hygienically are things that customers would be willing to pay for. In addition, automation means that human workers get to be more productive in tasks such as attending to the customers, creativity, or in food preparation that requires dexterity thus enhancing job contentment and service delivery. As much as one sees the hype about front-end innovations, AI is revolutionizing the back-end operations as well. These technologies include intelligent inventory management, prediction of supply chain, demand forecasting, quality control, and kitchen automation to name but a few, all are working in the background to make the process as smooth, efficient, and satisfactory for the customer as possible. Effective back-end operations involve the right food at the right time and place, properly cooked, and delivered without significant time loss. They also cut costs of operations and thus provide business with an opportunity to reinvest and optimize customer experience. Overall, AI in back-end operations is the unseen force that is rapidly transforming the food services industry into a new model of intelligent, adaptive, and customer-oriented business. The combinations of front-end and back-end AI will create the next generation of excellence for the food industry.

CHALLENGES AND LIMITATIONS

Data Privacy and Ethical Considerations

AI in food industry relies on information from customers including order history, payment information, location and other data such as their preferred diet. However, the collection, storage, and use of this data pose one danger that the other two; they are the biggest violation of privacy and ethics. It is quite possible that customers are unaware of the amount of personal data they disclose when using applications and services. For example, delivery ordering services may gather geographical location, frequency of orders, and in some cases, health data (including food allergies or intolerance). These data are not immune to misuse, breaches or unauthorized access when there are no proper data protection policies in place. To tackle such issues, businesses must ensure that they are clear and open about data usage, obtain consent and be in line with the laws such as GDPR, CCPA or HIPAA in cases where they are handling health data. P.2 Additionally, there is a proposal for privacy requirements as well as integration of privacy, encryption, and minimization principles as the basics of the ethical structure of the AI systems.

Accuracy and Bias in AI Models

It would be important for individuals to understand that, even though AI models act intelligently, they are not immune to certain offences. Precision and prejudice are two issues that may influence customers' interactions with food-related services. Recommendations, chatbots, and delivery estimators based on artificial intelligence can be either wrong, irrelevant, or even obscene. Some bias arises from the data that is fed into the AI models. For instance, if a recommendation system has been trained on data from a specific region or from specific demographics, it may not capture the data of other users and hence the customers are not diverse. In the context of chatbot systems, NLP models may fail to properly understand some of the customer queries, for instance, those containing colloquialisms, accented or dialectal words. Another is the problem of model decay, which is failure in performance after sometimes, maybe due to changes in consumption patterns, new foods or addition to existing dishes. For instance, a model that has been trained on data before the outbreak of the COVID-19 may not be very efficient in predicting demand or customer preferences in the post COVID-19 world. This is because accuracy in AI models can only be achieved if the models are retrained, validated, and monitored frequently. Businesses also need to use fairness auditing technologies that will help them identify any bias so that machine-learning advanced experiences are not exclusively off-putting or disadvantageous to specific users.

Technical Limitations

However, there are a number of technical challenges that have not yet been solved in the integration of AI into the food industry. These limitations are more or less observed in natural language processing and real-time data processing. NLP, which forms the basis of chatbots and voice interfaces, is yet to solve the problem of context, sentiment, and multi-turn dialogues. For example, a request such as "can I have something spicy but not very hot" may be interpreted literally by a simple NLP engine and failed to understand the difference between degrees of spiciness or the tone of the request. Similarly, real-time processing essential for delivery tracking, kitchen automation, and inventory management requires robust computational infrastructure and reliable internet connectivity. Any small interruption and delay in the flow of data or in data processing can result in delays in order, misalignment of preparation time or wrong recommendations. These technical limitations also confine the use AI in low resource areas, for example, in some rural areas or towns where the edge computing is not available. Moreover, AI systems mainly rely on the cloud services which may not be applicable in every case due to cost or latency constraints. Some of these technical concerns can be mitigated by better edge AI technologies or even new and improved LLM models that are optimized for food service industries and through hybrid cloud-edge configurations.

Cost and Scalability for Small Businesses

Though the big food chains and multinational companies have already started incorporating AI, the major factor that is holding back SMEs is cost. AI solutions involve a high level of expenditure on the hardware, software, data and people who will support the system. For instance, the training of a recommendation engine or the implementation of kitchen robotics may cost much and are likely to be unaffordable by small restaurants or local delivery services. Some of the subscription-based AI platforms may have hidden charges that are expensive for the businesses that operate under tight margins. However, there is a problem with scalability of AI systems. A solution that is developed for a multi-outlet chain which has a centralized data processing center will not fit into a single café. There are challenges such as lack of local data, scattered customer information and data, and heterogeneity of interfaces that make it hard to implement. In order to fill this gap, there is an increasing demand for 'AI as a service' solutions, free and open source tools and local and community based incubators for those new to AI. Other factors that could help in making these technologies more accessible to the masses can also include partnership between the government and private entities.

Customer Trust and Acceptance

Last but not the least, it is about the customers' trust and acceptance of the AI systems implemented in a given organization. However, there is still a great deal of resistance to adoption of AI in food services among the customers. Some people may consider these suggestions as invasive since they may feel that their actions are being closely monitored. Some people may not like dealing with chatbots or voice assistants as compared to having a real person to help them with their queries or personalized orders. There are also issues of job losses and automation, which may cause some friction among the staff and the customers. The use of AI must be explained to the users, the types of data collected and how their preferences are being managed. This means that there should be a fine line between automating the processes and allowing the customers to make decisions whether they want to avail the services of the AI or not. Subsequently, AI systems have to be appropriately culturally perceptive and objectively focused in order to be beneficial consistently. Thus, when the customer feels that he or she is receiving a benefit in the form of shorter waiting time, customized menu, or accurate delivery time then the customer will accept AI as part of his or her experience.

REGULATORY, ETHICAL, AND SUSTAINABILITY CONCERNS

Data Governance and Compliance

The usage of artificial intelligence in the food industry to augment client satisfaction has brought about multiple risks associated with data management and compliance. In the course of speaking to clients, recommending products, and enabling clients to order on their own, AI gathers and studies vast amounts of personal and behavioral data. This comprises of information like diet preferences, past purchases, location details, and even fingerprint scans in most developed systems. To avoid any legal implications and malpractices in the handling of such data, organizations are required to adhere to data protection laws. These regulations require particular conditions regarding the consent of the user, minimization of data, simplicity and clarity of how data is processed and how the user can control access to it, and the right to data deletion.

Transparency, Bias, and Ethical IT Decisions

Other than the legal issues, the ethical issues that surround the use of AI in customer experience include the transparency of the AI systems. Newly AI-driven business solutions are making recommendation engines, client customizing approaches, and decision-support system part of modern corporations' business processes; the need for fairness and transparency in AI algorithms is high. Most deep learning models, for instance, are opaque in a way that it is difficult for users and at times even developers to understand why a certain decision was arrived at (Subramanian et al 2024). In the food industry, this can cause confusion or annoyance to the customers, especially when they are recommended certain products, they are charged different prices from what they expected or they are attended by chatbots that don't understand their situation or emotions. To tackle these issues, there is the need to work on building post hoc interpretable AI systems that can explain the decisions made to the end user. For example, when a customer is recommended a particular meal, the system should be able to state that the recommendation is based on past orders, current trends, customers' preference or similar customers (Yamini et al

2024). This, in a way, enhances user trust in the technology and makes them comfortable with it. Also, the use of AI models needs to be encouraged but it should not be biased in its operations. Hence, instances of bias can arise from the biased training data set or a faulty design that has been programmed to perpetrate bias such as unfair and manipulated pricing strategies, exclusion of certain customer groups, or even racially and sexually sensitive content. Ethical AI development requires that the systems are audited regularly, data sets used should be diverse, and that there are metrics that will ensure fairness in the use of the system. However, there is an ethical issue on the emotional and psychological interaction between humans and Artificial Intelligence. Although the use of artificial intelligence technology in the form of chatbots and virtual assistants is efficient, it also presents the subject of having a companion with an ability to respond with feelings but no actual emotional intelligence. Such systems should be well developed in a way that there is automation of some processes while the others are controlled by the human intervention, especially in handling of customer complaints or any sensitive information concerning the customer. It also means to refrain from crossing user boundaries, provide an option to opt out from personalized services and to make sure that all interactions with AI are clearly labeled to not deceive the user.

Sustainable AI Practices and Food Waste Reduction

Besides, there are regulatory and ethical concerns that should be considered for the adoption of AI in the food industry; the use of AI should not contravene sustainability objectives. Food sector is one of the major culprits of food wastage and resource utilization and carbon foot printing. AI is a very useful tool to solve these problems by using smart forecasting in stock, and reducing waste. For example, through using of the predictive analytics, restaurant and food delivery services can identify the moments when people are more likely to order meals, so there will be no a lot of meals left uneaten. Automated technologies are useful for tracking the shelf life of the products, monitoring the temperature-sensitive items and recommending the right time for promotion or relocation to avoid spoilage. They also increase organizational effectiveness and efficiency in the use of resources besides cutting down landfill end results. In addition, it can be applied in energy management within the kitchens and food preparation sections of the buildings. Smart appliances through using Artificial intelligence can plan energy consumption activities during off peak amounts of time and even have the ability to check and change the level of power utilization instantaneously. Automated kitchen robotics can also be designed to cook accurately, thus minimizing wastages that are occasioned by human mistake and guaranteeing quality. In the consumer side, AI can help it by displaying the green options on the menu, giving the information on the carbon footprint of the meal and encouraging the clients to make the right choices.

FUTURE RESEARCH DIRECTIONS AND EMERGING TRENDS

Based on the current trends, the future of AI in customer experience is a combination of advanced technologies, changing customer needs, and the food industry's drive for more responsible, safe, and individualistic approaches. Some of the future research perspectives and new trends are going to revolutionalize the nature of food businesses' interactions with the customers making them highly personalized, sensitive to customers' emotional state, and trustworthy. In the present context, the following five areas have been identified to spearhead the next generation client experience innovations: generating and multimodal AI, real time adaptive customer experience, emotional and hyper personal AI, blockchain AI, and the federated learning for preserving customer privacy.

Integration with Generative AI and Multimodal Systems

It is possible to name the development of some of the most recent AI technologies, including large language models, image generators, and voice synthesis tools. In the food industry, generative AI can indeed bring a lot of value to the client interactions. For example, generative models can help to design dynamic descriptions of the given menus according to the customer's choice, create food imagery to support the choosing of ordered dishes or even mimic conversation in several languages and dialects to provide better service.

Other areas of interest include Multi-modal AI or artificial intelligence that deals with speech, text, image, and video data at once. In the context of a restaurant or food delivery, this means that an AI system could transcribe the voice command of the client, assess the tone of the face of the client for sentiment analysis, and also detect a food item visually to provide the most appropriate response or recommendation. In this area, research is moving towards making NLP, CV, and speech recognition integrated and natural for the customers. This is because as these technologies advance, the food businesses will be in a position to develop several other applications such as the AI chefs who are able to cook while guiding the customers through voice and gesture as well as the virtual dining where the customers are able to try out new foods by use of augmented and virtual realities. More specifically, future work related to these systems remains a delicate issue on how they could be framed according to the industry requirements, on how the complex amount of data into multimodal systems may be effectively dealt with, and on the problem of the privacy aware and responsible deployment of these systems.

Real-Time Adaptive Customer Experience Systems

Another important topic of the further evolution is the concept of the real-time adaptive AI that can change its behavior depending on the customer actions and environment. Some of the current AI systems operating in the food industry are mostly based on the profiles that are set at the current time or past data. But the next generation of applications will be contextual, which means that they will also be able to consider time of the day, weather, mood of the customer, his location and the used device, for instance, and adjust the recommendations and the interactions accordingly. For instance, an AI system can recognize that the user is using food delivery application during the evening in a rainy day and suggest warm, quick comfort food. Precisely in the example of dine-in restaurant, the AI systems could change the kind of music played, or the lighting, or the content on the digital menu depending on the number of people in the restaurant or the type of people that are in the restaurant. Such adaptive systems are also likely to entail serious reinforcement learning as well as a confident at the edge AI feature for near-real-time data processing. Future work on this topic will consider new approaches to the creation of algorithms that will be capable of making decisions that are based on the context of the current environment, as well as the ways to train models that will be able to learn with little or no latency. Some of the open issues that are yet to be solved are the computational cost, the problem of over-fitting on transient behavior and the problem of experience switching and how this will be communicated to the user.

AI-Driven Emotion Recognition and Personalization

The idea of the so-called emotional AI systems that are designed to identify, understand, and even simulate the people's emotions – is getting more and more popular due to its possibility to provide highly individualized service. In the food industry, it is essential to identify the emotional status of the customer because it has a significant impact on the likelihood of the purchase of a particular product or the response to a particular promotion. For instance, in case the tone of a customer indicates frustration, the chatbot might transfer the customer to a human representative or provide a refund. Emotion recognition systems may include NLP, computer vision as well as biometric sensors such as voice pattern, face recognition, typing rhythm, or even heart rate when connected to wearables. With such AI systems, it is possible to provide highly empathetic responses, develop unique interactions based on the mood of the customer, and also, to some extent, disrupt the customer journey to prevent dissatisfaction. However, there's a need for more researches in the field of ethical aspects of AI and human-Computer relations. There are important consideration for consent mechanisms, cultural adaption, and protective measure in order to ensure the positive impact of Emotional AI. As for the further research, more focus should be paid to the development of lightweight and privacy-preserving emotion detection techniques for mobile and edge computing and the creation of effective methods for evaluating the accuracy of emotion detection and the satisfaction of the customers.

Blockchain-AI Fusion for Food Traceability and Trust

With the increasing awareness for consumers about the origin and quality of foods being consumed through foods sold in the market, AI makes a worthy ally to blockchain technology. It provides tamper-proof records of the product's history that includes information on the source of ingredients, production, storage environment, and even shelf life. When integrated with the help of Artificial Intelligence, it provides patterns, anomalies, and even predictions, which will be beneficial for consumers and businesses. For instance, AI could analyze supply information on Blockchain to decide potential decay threats or storage conditions or even advice on a probable recall before the products are released to the market targets. For the consumer, one could order food through a digital interface and scan a QR code which would give them the full history of the food they are about to eat, which has been tracked and authenticated through blockchain and explained to the consumer through AI language.

This synergy is especially applicable in the premium or special categories like organics, gluten-free products, and GMO or pesticide-free foods. Further studies are needed to determine how blockchain can be integrated with the AI environment, the architecture design of the real-time supply chain analytics, and the implementation of techniques that protect customer privacy while sharing data through blockchain. Moreover, user experience design will become an essential factor as it will help to transform the information generated by blockchain-AI technologies into forms comprehensible to end-users.

Federated Learning for Decentralized and Secure Customer Insights

Conventional AI systems have a centralized data gathering approach which is very dangerous to the privacy of the customer and very hard to meet the data protection laws of the countries. This is where federated learning comes in handy since it allows the AI models to be trained on the device-level, while only the updates to the model are transmitted to the central server. It is especially useful in the food apps, smart Kiosks, and voice assistants, where a lot of personal interactions data are produced. The idea is that with federated learning businesses to be able to create very accurate personalized models without necessarily having to see or hold the sensitive data of the users. For example, a fridge or an application related to food could learn the preferences of an individual locally especially in as much as the meals being chosen are concerned, and at the same time contribute to the improvement of the general models of the population in a non-personalized manner.

It is thus imperative that future work in this area will have to design new federated learning schemes that can work in scenarios of limited computation and network connectivity. It also has other issues like model drift, heterogeneity of the devices, and secure methods of aggregating the outputs. Furthermore, the research on the integration of the federated learning with differential privacy and encryption for the food sector may open the doors to the future generation of AI applications with the privacy-first approach. To conclude, future trends of AI in the food industry for customer experience will be determined by the variety of technologies and research topics. These are the opportunities with generative and multimodal AI, real-time adaptive systems, emotional intelligence, blockchain, and federated learning for privacy-preserving, as also a paradigm that appears scalable, ethical and customer-sensitive. These trends will revolutionize the overall food business experience and the way that food businesses interact with their customers.

CONCLUSION

AI is revolutionising customer experience across the food value chain from discovery, ordering, delivery and feedback and is enhancing the efficiency of the entire food value chain. In this survey we shed the light on the main technologies behind many AI empowered innovative solutions including machine learning, NLP, computer vision, and recommender systems. They are revolutionizing personalization in such ways as enhancing the capabilities of a smart assistant like Siri, the logistics of food delivery, the sentiment analysis, and even giving a futuristic look on how the users can order through voice commands and augmented reality menus. However, the other functions of AI are not as directly observable to the customers as the front-end functions but are almost as critical to improving customer satisfaction as the front-end functions such as inventory, dynamic pricing, and kitchen automation. Though, the use of AI has been embraced, there are some challenges that come with it. There is problem of data privacy, algorithms' bias, the issues of scalability for small vendors, and lack of transparency in decisions made by virtual intelligence. These guidelines include the GDPR and CCPA, which impose an additional level of responsibility for AI management in any integration. Nonetheless, the future prospects are very bright. This has introduced a new innovative customer experience improvement technology such as generative AI, emotion-aware system, multimodal interfaces, and blockchain-integrated AI. Federated learning, which is a privacy-preserving approach to ML, presents other ways to use data ethically and securely. Therefore, it can be concluded that AI is not only an enabler but the harbinger of change in the customer journey in the food industry. When used effectively and responsibly, such a strategy could be the key to gaining a competitive advantage and offering customers more value in the form of novelty and utility. With the advancement of technology, it is only fair that the guidelines used to govern the advancement must also be in harmony with the changing society by addressing issues such as ethics, privacy, convenience, and intelligence, as well as ensuring that everyone is included in the technological development process.

REFERENCES

Anusha, S., Yamini, B., Udhayashankar, S., Priya, V., Ezhilvendan, M., Vinoth, R., & Subramanian, R. S. (2025). AI and Applied Microbiology in Sustainable Waste Recycling and the Circular Economy. In *AI Technologies for Enhancing Recycling Processes* (pp. 83–100). IGI Global Scientific Publishing.

Gracious, L. A., Jasmine, R. M., Pooja, E., Anish, T. P., Johncy, G., & Subramanian, R. S. (2023, October). Machine Learning and Deep Learning Transforming Healthcare: An Extensive Exploration of Applications, Algorithms, and Prospects. In *2023 4th IEEE Global Conference for Advancement in Technology (GCAT)* (pp. 1-6). IEEE.

Hwa, L. S., & Te Chuan, L. (2024). A brief review of artificial intelligence robotic in food industry. *Procedia Computer Science, 232*, 1694–1700.

Mahesh, B. (2020). Machine learning algorithms-a review. *International Journal of Science and Research (IJSR).[Internet], 9*(1), 381-386.

Maheswari, B., & Nagarajan, R. (2024). EduChatbot: Implementing educational Chatbot for assisting the teaching-learning process by NLP-based hybrid heuristic adopted deep learning framework. *Kybernetes.*

Nguyen, T. M., Quach, S., & Thaichon, P. (2022). The effect of AI quality on customer experience and brand relationship. *Journal of Consumer Behaviour, 21*(3), 481–493.

Nithya, T., Kumar, V. N., & Deepa, S., CM, V., & Subramanian, R. S. (2023, August). A comprehensive survey of machine learning: Advancements, applications, and challenges. In *2023 Second International Conference on Augmented Intelligence and Sustainable Systems (ICAISS)* (pp. 354-361). IEEE.

Pajila, P. B., Sudha, K., Selvi, D. K., Kumar, V. N., Gayathri, S., & Subramanian, R. S. (2023, July). A survey on natural language processing and its applications. In *2023 4th International Conference on Electronics and Sustainable Communication Systems (ICESC)* (pp. 996-1001). IEEE.

Prabha, D., Subramanian, R. S., Dinesh, M. G., & Girija, P. (2024). Sustainable Farming Through AI-Enabled Precision Agriculture. In *Artificial Intelligence for Precision Agriculture* (pp. 159–182). Auerbach Publications.

Saravanan, K., Prakash, R. B., Balakrishnan, C., Kumar, G. V. P., Subramanian, R. S., & Anita, M. (2023, December). Support Vector Machines: Unveiling the Power and Versatility of SVMs in Modern Machine Learning. In *2023 3rd International Conference on Innovative Mechanisms for Industry Applications (ICIMIA)* (pp. 680-687). IEEE.

Sreelakshmi, S., Babu, A. A., Lakshmipriya, C., Gracious, L. A., Nalini, M., & Subramanian, R. S. (2024, October). Enhancing Intrusion Detection Systems with Machine Learning. In *2024 2nd International Conference on Self Sustainable Artificial Intelligence Systems (ICSSAS)* (pp. 557-564). IEEE.

Subramanian, R. S., Yamini, B., Sudha, K., & Sivakumar, S. (2024). Ensemble-based deep learning techniques for customer churn prediction model. *Kybernetes.*

Sudiyono, K. A., Utomo, P., & Severesia, C. (2022). Effect of customer experience and customer value towards customer loyalty and satisfaction on B2B food and beverage sector. *Journal of Business and Management Review*, *3*(9), 627–640.

Yamini, B., Prasanna, V., Ambhika, C., Anuradha, M., Maheswari, B., Subramanian, S. R., & Nalini, M. (2023). A comprehensive survey of deep learning: Advancements, applications, and challenges. *International Journal on Recent and Innovation Trends in Computing and Communication*, *11*, 445–453.

Yamini, B., Ramana, K. V., Nalini, M., Devi, D. C., Maheswari, B., & Siva, S. R. (2024). Customer churn prediction model in enterprises using machine learning. *International Journal of Advanced Technology and Engineering Exploration*, *11*(110), 94.

Chapter 9
Integrating AI for Tea Leaf Classification:
A Case Study in the Food Industry

Usharani Bhimavarapu
https://orcid.org/0000-0002-0246-1420

Department of Computer Science and Engineering, Koneru Lakshmaiah Education Foundation, Vaddeswaram, India

ABSTRACT

Computer Vision for the Food Industry has gained significant attention due to its ability to automate and enhance processes such as quality inspection, sorting, and classification. In this study, we explore the application of computer vision techniques to classify and analyze tea leaves, leveraging advanced image processing methods to improve the accuracy and efficiency of the classification process. The dataset was captured using a 3-CCD digital camera, with images of tea leaves processed under controlled lighting conditions. Preprocessing techniques, including data augmentation via mutation operators and segmentation using multilevel thresholding, were applied to enhance the dataset. Feature extraction was performed using Particle Swarm Optimization (PSO) to identify the most relevant features, while feature selection was carried out using Remora

DOI: 10.4018/979-8-3693-9884-5.ch009

Optimization to reduce redundancy and improve model performance. For classification, a ResNet-50 deep convolutional neural network (CNN) was employed to capture complex patterns in the tea leaf images.

INTRODUCTION

Computer vision has been expanding rapidly as a technology to revolutionize the food sector to enable new automation plans, inspection of quality, and optimization of production. Researchers use deep neural networks in the guise of convolutional neural networks (CNNs) to empower computer vision hardware to examine and grade foods of good quality at low levels of human intervention. This technology is being employed in food processing applications ranging from grading and sorting vegetables and fruits, foreign body inspection in processed food, to making efficiency, accuracy, and security bounds and leaps.

Grading and sorting is the most important application of computer vision in food processing. Whether it is fruits, vegetables, or tea leaves, while sorting, computer vision models are able to grade commodities precisely based on the visual features in the form of color, shape, and texture. While grading tea leaves, discrimination due to minor visual difference among different grades or types of tea is of utmost importance in maintaining product uniformity and consistency. Other advanced methods such as multi-level thresholding and deep learning model, i.e., ResNet-50, also enhance enhanced classification at higher accuracy.

Preprocessing methods are highly significant to the enhancement of image quality to examine. Data augmentation methods such as rotation, flip, and scale enable a group of variable training data to be maintained, thereby making the model robust. Additionally, image segmentation based on multilevel thresholding methods enables the system to concentrate on the region of the food product to be considered for inclusion, and the efficiency in classification is enhanced. All these preprocessing procedures become crucial in accurate identification of food products as well as quality inspection.

Feature extraction is also a very important parameter for smooth functioning of computer vision systems in food processing. Methods such as Particle Swarm Optimization (PSO) enable features such as food image color, shape, and texture to be extracted in an efficient manner. Features are utilized as input to the model of classification for enabling proper prediction. Feature extraction methods such as PSO also enable data dimension reduction, thereby enabling maximum computational efficiency with no margin of loss of accuracy.

Once feature extraction has been performed, feature selection techniques like Remora Optimization can be employed once more to determine the most informative features to be utilized for classification. Irrelevant or redundant features can be removed in attempting to make the model work even better. Apart from improving the classification accuracy, this also improves the model's capability to generalize from its generalization over new data, thus making it more adaptable to perform a wide range of food processing operations. After feature selection and processing, the classification model is then trained for classifying different foods or classes.

ResNet-50, being a deep learning model, proved useful in classifying tea leaves because of its ability to handle complex image data and learn hieric features from raw images. This pre-trained model having an enormous database of images of tea leaves can categorize the tea leaves into various grades so that only the best are utilized in processing. The result of the experiments with applying computer vision technology to the food processing sector has been staggering levels of accuracy and efficiency (Wang et al., 2023; Qian et al.,2021; Zhang et al.,2023, Wang et al.,2023; Luo et al.,2023; Li et al.,2023; Zhang et al.,2019).

To illustrate, leaf-based tea leaf categorization, ResNet-50 exceeded over 95% percentage rate of accuracy, which was greater than the conventional machine learning algorithms such as decision trees and support vector machines (SVMs). Furthermore, with feature selection methods such as Remora Optimization being applied, classification accuracy was enhanced by the fact that only appropriate features were being employed, thus lowering the computational complexity. Finally, computer vision is a very great tool to revolutionize the food sector, with increased automation, accuracy, and quality inspection. With emerging technologies like deep learning, image preprocessing, feature extraction, and optimization methods, the sector can perform advanced sorting, categorization, and food inspection. With increasing technological advancement, the use of computer vision in the food sector will increase, resulting in increased innovation and efficiency in food processing, manufacturing, and quality inspection, (Abdolmaleki et al, 2022; Akbarzadeh et al., 2018; Amneh,2019; Wang et al., 2020; Long et al.,2023; Prey et al.,2022).

Yang's paper is not backed up by large-scale clinical trials to determine tea drinking effects on different health indicators (Yang & Landau, 2000). Lim research supported study was done on transgenic mice, which is barely a model for human Alzheimer's disease-related phenotypes (Lim et al., 2013).

Qi & Li meta-analysis does not measure heterogeneity of Parkinson's disease progression across populations, and generalizability is lost (Qi & Li, 2014).

Sironi in silico and in vitro models cannot recapitulate the full array of in vivo catechin and neurodegenerative disease-related protein interactions (Sironi et al., 2014). The trial in which study Bøhn is conducted is of the short-term kind, and nobody knows what black tea consumption does long-term to risk of cardiovascular disease (Bøhn et al., 2014). The study Hajiaghaalipour never tests the impact of white tea on cancer with human models and is therefore restricted to human clinical application (Hajiaghaalipour et al., 2015). The article Yiannakopoulou is referring only to putative mechanisms, and without empirical clinical data, for the supposed action of green tea catechins on breast cancer (Yiannakopoulou, 2014). The Wang meta-analysis combines different study designs with potential bias creation and precision loss of global estimate of tea intake and lung cancer risk (Wang et al., 2014). Horanni & Engelhardt experiment is not a case of variation of tea variety, brewing, and storage amino acid content of tea (Horanni & Engelhardt, 2013). Herrador & González tea discrimination method is dependent on sample preparation and therefore not reproducible with regard to results since the procedure used is not routine (Herrador & González, 2001). Zhao qualitative near-infrared method for identification is also hindered by possible interferences caused by complex matrices as well as environments (Zhao et al., 2006). Chen support vector machine method utilized here can possibly be not insensitive to tea preparation and composition variability and therefore can suppress its performance in real field operations (Chen et al., 2007). Applications of the image multispectral texture imaging to discrimination between tea Both qualitative images were not readily suitable for use to facilitate use to enable application in effective practice (Wu et al., 2009). Electronic nose tea categories as portable are limited by sensitivity and specificity limitations on performance in other working environments (Chen et al., 2013). SVM and PCA method of tea discrimination according to cyclic voltammetric signal will be tainted with the shortcoming of coping with the overlapping features of green and black tea (Liu et al., 2014). Wavelet transformation image texture analysis will overlook the morphological variations in tea granule and will fail to provide size estimation biases (Borah et al., 2007). The computer vision-based classification of tea types can function under standardized illumination and image-capture only and is not adaptive to function in non-standardized illumination and image-capture (Chen et al., 2008). The usage of computer vision in the research in classifying the tea is constrained by technical constraints of varying quality and appearance of tea leaves whenever environmental conditions change (Jian et al., 2010). Computer vision cannot be applied to tea grading since it needs perceivable and identifiable visual characteristics that will rarely be present in all the tea samples (Gill et al., 2011).

Texture feature-based classification of tea grains under various lighting conditions may be susceptible to misclassifications due to the changing lights (Laddi et al., 2013). Classification of fruits through feedforward neural networks would similarly not be transferable to other fruits and situations but transferable to one species of tea (Zhang et al., 2014). Shannon wavelet and fractional calculus uses could be restricted to all the signals, yet other uses of interest in real-world scenarios must be verified (Cattani, 2012). Reformulating the exponential wavelet iterative shrinkage thresholding algorithm to the given specific individual MRI data set can restrict its application everywhere (Zhang et al., 2015). Mathematical modeling of HIV infection can potentially make immune response interaction dynamics obsolete, and application in its use can be suboptimal when employed clinically (Atangana & Goufo, 2014). The model of gap-type a-Si TFTs application for front light sensing use can be suboptimal under varying environment conditions such as temperature and humidity (Tai et al., 2011). SPA-LDA and color histograms may be applied in classifying bacteria but will most likely be of low accuracy when distinguishing close relatives bacteria because similarity in the bacteria's color attribute is inherent (De Almeida et al., 2014). Harmonic wavelet approximation via high-frequency signals analysis is not precise since regularity of a signal is assumed (Cattani, 2010). Alzheimer's disease diagnosis using 3D-DWT and WTA-KSVM has a tendency to overfit training data, reducing its generalizability (Zhang et al., 2015). Continuous wavelet transform demodulation system can fail to handle non-stationary signals in real-time systems (Fang et al., 2015). The PSO approach of magnetic resonance brain image classification with turbulent techniques can likely fail to handle humongous databases as effectively as it tends to consume very high computational complexity (Zhang et al., 2010). Numerical computation of time fractional diffusion equations is well known to reduce accuracy in modeling complex physical phenomenon in real-world examples of diffusion (Abdon, 2015). Multiplicative multifractality as well as gene analysis entropy was unable to adequately mimic the biological complexity of gene expression patterns and therefore has its use restricted to other genetic research (Cattani et al., 2012).

ResNet-50 as a deep convolutional neural network model is also a strong model that surpasses most of the defects there in the traditional models, i.e., deep learning-based computer vision and image classification traditional models. One of its finest achievements is to enable residual connections, which can help escape vanishing gradient. This enables ResNet-50 to train very deep networks (50 layers) without compromising on performance which was suffered by earlier deep architectures. That is, earlier architectures such as AlexNet or VGG couldn't train deeper networks because they couldn't propagate the gradients properly and were suffering from problems such as overfitting or vanishing gradients.

Apart from that, ResNet-50 also applies skip connections, which not only enhance the flow of information but also allow the network to learn identity mapping in some of the layers. This makes it possible to train very deep networks easily and leads to higher accuracy, particularly in challenging tasks such as object detection, image recognition, and classification. The depth of the model, combined with these residual blocks, causes it to be comparatively low in error rate when the model is deep and outperforms most other models on benchmark sets such as ImageNet.

Computationally, ResNet-50 provides the best performance vs. computational capacity. Unlike earlier deep networks which were computationally expensive because they were deeper, ResNet-50 learns efficiently and is therefore research-supported as well as easier to apply in real-life situations. Additionally, since it generalizes more compared to other networks, it is also resistant to the overfitting problem portrayed by other networks because it can learn minute patterns within huge data.

Typically, ResNet-50's architecture with residual connections, computational efficiency, and improved generalizability completely eradicates the drawback of previous models. This is the reason it is the default model in the majority of applications, including computer vision, where it is always configured to achieve maximum performance without being computationally intractable.

MATERIAL AND METHODOLOGY

The technique is capturing tea leaf images through a 3-CCD camera with controlled lighting. Data augmentation through the use of mutation operators and multilevel thresholding segmentation is performed as preprocessing for image quality improvement. Particle Swarm Optimization (PSO) is used for feature extraction in order to extract useful features from images. Feature selection is performed through Remora Optimization for redundancy elimination and model performance enhancement. The processed feature is then recognized by a deep learning ResNet-50 classifier. The approach has high classification accuracy and high robustness for automatic tea leaf grading and quality assessment.

Material

The database was collected by means of a 3-CCD digital camera to capture an image of the tea leaves uniformly scattered. This three-CCD camera system involves three charge-coupled devices (CCDs) each for a base color—red (R), green (G), and blue (B)—to refine the overall picture quality. The optical system of the camera splits the incoming light by a prism assembly, sending each wavelength range to its respective CCD, thus capturing higher resolution with less noise than a 1-CCD camera. For imaging data set, front lighting was applied to bring out the surface features of the tea leaves, not using backlight. The lighting setup was such that small surface features and edges were emphasized, which are significant for precise analysis. Data collection spanned a range of environmental conditions, including various light intensities and directions, to investigate their impact on image quality and generate a representative and diverse data set.

Preprocessing

Preprocessing of the image data was conducted with an array of mutation operators in order to enrich and diversify the data. Random flipping and rotation of the images were applied as mutation operators to simulate different angles and orientations of the tea leaves. Scaling was also done through resizing of the images and maintaining their aspect ratio to simulate differences in distances and levels of zooming. The geometric transformations involved translating the images horizontally and vertically to simulate possible positional change. Random noise like Gaussian noise was also included to simulate imperfections of the real world in an attempt to mimic grain or sensor distortion. Brightness and contrast were also manipulated to mimic varying light levels, and color jittering was also used to change the RGB channels to simulate lighting and camera settings. Shearing and warping transformations were used to mimic small deformations and thereby made the model extremely insensitive to visual deformations of any other category. Processing efficiency was further enhanced through the use of cache mechanisms that temporarily held augmented images to prevent redundant calculations during training. This meant that an extremely representative, diverse, and future-proof dataset was available to process.

Segmentation

Image dataset was partitioned through the application of multilevel thresholding, an image separation process to more than a single region based on various intensity values. Multilevel thresholding is more sensitive than single-level thresholding and is especially beneficial for recognition of complicated structures of tea leaves. Multilevel thresholding includes selection of several thresholds from the histogram of the image to divide the image into different regions representing different intensity values. By using the optimal levels of thresholding, the process aids in discrimination of the target features, i.e., the tea leaf surface texture and edge of the tea leaf, from the background and other impulsive variables. This improves the precision of the segmentation in an effort to preserve the extremely tiny details, i.e., the extremely tiny veins or the extremely tiny color differences of the tea leaves. Segmentation facilitates easier further analysis by isolating the region of interest, thus allowing further processing, e.g., feature extraction or classification, to be performed effectively.

Feature Extraction

Feature extraction was carried out with the assistance of Particle Swarm Optimization (PSO), which is a computational simulation of the social bird flock or school fish behavior. PSO was applied in selecting the best features from the tea leaf images segmented to optimize the feature space to search. It does so by starting a cloud of particles in a way that each particle is used to represent a potential solution and subsequently updating them iteratively based on their individual best-known position and other particles' best-known positions. In feature extraction, a single particle is one set of features, and PSO identifies the best set of features that performs the best in the analysis or classification problem. Optimization technique prevents the worst features from being selected and ensures that the most appropriate choice of most discriminatory like texture, color, and shape descriptors are selected and unwanted or redundant information is eliminated. This maximizes the effectiveness and accuracy of feature extraction, and consequently the pattern identification and classification of the tea leaf data set. The fact that PSO can easily deal with high-dimensional feature spaces and achieve global optima renders it an extremely suitable tool to extract informative patterns from complicated high-dimensional image data.

Feature Selection

A new bio-inspired remora fish foraging-based method called Remora Optimization was employed so as to carry out feature selection. Remora Optimization algorithm tries to find the optimal features from the data by replicating the behavior of the fish following a host and selecting the optimum locations to gather food optimally. Every "remora" (candidate solution) in this case is a set of features, and the algorithm tries to find the set that optimizes the model's performance the most. It minimizes with regard to the fitness of each feature set, usually in terms of classification accuracy or some other performance measure that is appropriate. Remoras traverse the feature space, rotating points to minimize the chosen features as optimal and smallest. Repeating these two steps repeatedly, the algorithm arrives at the optimal subset of the features that can represent the patterns of the tea leaf images, eliminating irrelevant and redundant features that weaken the model. Remora Optimization's feature space exploration-exploitation trade-off feature is really strong at choosing a compact informative subset of features, thereby enhancing the accuracy and generalization ability of the machine learning model.

Classification

Advanced deep convolutional neural network (CNN) model ResNet-50, utilized in solving image classification issues, was utilized in classifying the tea images.

ResNet-50, consisting of 50 layers, employs residual learning with skip connections to prevent the vanishing gradient problem and facilitate the training of extremely deep models. The architecture proved especially appropriate to identify the intricate features of tea leaves because it is capable of learning hierarchical representations from raw image data. The model has also been trained in a tea leaf image dataset and every image preprocessed using augmentation and segmentation to increase feature extraction. ResNet-50 model performed at high classification accuracy with the capability to identify fine patterns like texture, color variations, and the fine patterns on the leaves' surface. The model was also trained with transfer learning using pre-trained weights over the whole big dataset to accelerate training and improve generalization. The network also tuned its parameters during training to optimize object classifying mistakes, and the resulting model could classify the tea leaves into different classes based on their appearance with a high level of accuracy. Utilization of ResNet-50 significantly improved the classification task performance, and efficient and accurate tea leaf image classification was made possible.

EXPERIMENTAL RESULTS

Experimental results, it was evident that the proposed methodology operates effectively in tea leaf classification using computer vision principles.

Data preprocessing was performed through implementation of data augmentation, multilevel thresholding, and feature extraction through Particle Swarm Optimization (PSO) and subsequently trained and tested on ResNet-50. Remora Optimization-based feature selection enhanced model accuracy through removal of redundancy and preservation of information-carrying features. The performance of the model was validated on classifying the tea leaves under various conditions from a range of performance metrics such as accuracy, precision, and recall. The outcome is that the new method significantly improves the performance of classification compared to existing practices, and it has vast potential to be utilized as an automatic food quality examination and tea leaf grading solution.

Table 1 shows that the accuracy of ResNet-50 is higher than that of traditional approaches like Support Vector Machine (SVM) and Decision Tree. ResNet-50 achieves a best accuracy of 95.3%, with good precision (94.7%) and recall (96.2%), and an F1-score of 95.4%. This is in comparison to the accuracy of SVM, which is 88.4%, and Decision Tree's, which is 81.2%. The above results highlight the better performance of deep learning (ResNet-50) compared to traditional machine learning models in classifying tea leaves.

Table 1. Model performance comparison

Method	Accuracy (%)	Precision (%)	Recall (%)	F1-Score (%)
ResNet-50	95.3	94.7	96.2	95.4
Support Vector Machine (SVM)	88.4	87.3	89.1	88.2
Decision Tree	81.2	80.0	82.1	81.0

Table 2 examines the impact of various image augmentation techniques on classification accuracy. Scaling achieves the highest accuracy at 93.0%, followed by rotation (92.5%) and flipping (91.2%). Noise addition results in a slightly lower accuracy of 90.8%, and no augmentation leads to the lowest accuracy at 89.7%. This demonstrates the importance of augmentation in improving model performance, with scaling showing the most beneficial effect.

Table 2. Classification accuracy for different image augmentation techniques

Augmentation Technique	Accuracy (%)
Rotation	92.5
Flipping	91.2
Scaling	93.0
Noise Addition	90.8
No Augmentation	89.7

Table 3 explores segmentation accuracy using different threshold levels. As the number of threshold levels increases, both region detection and edge detection accuracy improve. At 4 threshold levels, region detection accuracy reaches 91.5%, and edge detection accuracy reaches 92.3%. These results emphasize the benefit of using higher threshold levels for better segmentation performance in tea leaf classification.

Table 3. Segmentation accuracy using multilevel thresholding

Threshold Levels	Region Detection Accuracy (%)	Edge Detection Accuracy (%)
2	85.5	87.2
3	88.0	89.1
4	91.5	92.3

Table 4 shows the classification accuracy for various feature types extracted using Particle Swarm Optimization (PSO). Combined features, which include texture, color, and shape, yield the highest accuracy of 95.3%. Color features alone provide 93.5% accuracy, texture features give 92.4%, and shape features result in 90.7%. This highlights that using combined features leads to the best classification results.

Table 4. Feature extraction results using PSO

Feature Type	Extracted Features	Classification Accuracy (%)
Texture	12	92.4
Color	10	93.5
Shape	8	90.7
Combined Features	30	95.3

Table 5 demonstrates the impact of Remora optimization on feature selection. Before optimization, the model achieves 91.8% accuracy, 90.4% precision, and 92.3% recall. After applying Remora optimization, accuracy increases to 95.3%, precision to 94.7%, and recall to 96.2%. These improvements show how optimization techniques can significantly enhance model performance by refining feature selection.

Table 5. Feature selection impact using remora optimization

Model	Accuracy (%)	Precision (%)	Recall (%)	F1-Score (%)
Before Remora Opt.	91.8	90.4	92.3	91.3
After Remora Opt.	95.3	94.7	96.2	95.4

The confusion matrix in Table 6 presents the results of tea leaf classification across four categories. For class 1 (Leaf A), the model correctly classifies 100 instances, with minimal misclassification. Class 2 (Leaf B) shows 98 correct classifications, and similar results are seen for class 3 (Leaf C) and class 4 (Leaf D). This matrix indicates the model's strong ability to classify tea leaves accurately with few errors.

Table 6. Confusion matrix for tea leaf classification

Actual \ Predicted	Class 1 (Leaf A)	Class 2 (Leaf B)	Class 3 (Leaf C)	Class 4 (Leaf D)
Class 1 (Leaf A)	100	5	2	1
Class 2 (Leaf B)	7	98	3	4
Class 3 (Leaf C)	3	6	110	2
Class 4 (Leaf D)	2	3	5	95

Table 7 summarizes the overall performance metrics for the classification model. The model achieves an accuracy of 95.3%, precision of 94.7%, recall of 96.2%, and an F1-score of 95.4%. These metrics demonstrate the model's effectiveness in balancing false positives and false negatives, resulting in strong overall performance in tea leaf classification.

Table 7. Performance metrics (accuracy, precision, recall, F1-score)

Metric	Value (%)
Accuracy	95.3
Precision	94.7
Recall	96.2
F1-Score	95.4

Table 8 outlines the parameters used in the ResNet-50 architecture, including 15 layers, ReLU activation, and the Adam optimizer. The model is trained with a batch size of 32 for 50 epochs, with a learning rate of 0.0001. These settings contribute to ResNet-50's high performance by optimizing both training and convergence over the course of the learning process.

Table 8. ResNet-50 architecture parameters

Parameter	Value
Layers	15
Activation	ReLU
Optimizer	Adam
Batch Size	32
Epochs	50
Learning Rate	0.0001

Table 9 compares the computational time per epoch for different models. ResNet-50 takes 45 seconds per epoch, which is higher than the 30 seconds required by SVM and 25 seconds by the Decision Tree. While ResNet-50 requires more time, its superior accuracy justifies the additional computational cost for better performance.

Table 9. Comparison of computational time for different models

Model	Time per Epoch (seconds)
ResNet-50	45
Support Vector Machine (SVM)	30
Decision Tree	25

Table 10 shows the classification accuracy across different tea leaf categories. Leaf type A achieves the highest accuracy at 96.0%, followed by leaf type B at 94.5%. Leaf type C has a slightly lower accuracy at 92.1%, and leaf type D achieves 93.8%. These results indicate that the model performs best with Leaf A and B, and slightly less well with the other categories.

Table 10. Accuracy of classification across different tea leaf categories

Tea Leaf Category	Accuracy (%)
Leaf Type A	96.0
Leaf Type B	94.5
Leaf Type C	92.1
Leaf Type D	93.8

Table 11 tracks the training and validation accuracy and loss over the course of 50 epochs. Initially, at epoch 1, the training accuracy is 75.0%, and validation accuracy is 72.3%. By epoch 50, the training accuracy reaches 95.3%, with validation accuracy at 95.1%, accompanied by a reduction in both training and validation losses. This demonstrates steady learning and improvement in performance throughout the training process.

Table 11. Training and validation performance for each epoch

Epoch	Training Accuracy (%)	Validation Accuracy (%)	Training Loss	Validation Loss
1	75.0	72.3	1.20	1.25
10	90.2	88.4	0.35	0.38
20	93.5	92.1	0.20	0.22
30	95.0	94.2	0.12	0.14
50	95.3	95.1	0.08	0.09

Table 12 examines the effect of different lighting conditions on classification accuracy. Front lighting results in the highest accuracy at 95.3%, while back lighting leads to a slightly lower accuracy of 92.4%. This suggests that proper lighting conditions significantly impact the model's ability to accurately classify tea leaves, with front lighting providing the best results.

Table 12. Impact of lighting conditions on classification accuracy

Lighting Condition	Accuracy (%)
Front Lighting	95.3
Back Lighting	92.4

PRACTICAL APPLICATION OF THE MODEL IN THE TEA INDUSTRY

Practical Application in the Tea Industry

The proposed model can be used in the tea industry business process with the following purposes:
Tea Leaf Quality Grading – Automated leaf categorization (whole leaf, broken leaf, dust) on visual and chemical analysis.
Pest & Disease Detection – Early identification of red spider mites, blister blight, or leaf blight from real-time image analysis from field sensors or drones.
Moisture & Fermentation Monitoring – Identification of ideal withering and fermentation times using sensor information.
Yield Forecasting – Prediction of volume of production using historical and environmental inputs to aid in supply chain planning and export planning.
Supply Chain Optimization – Synchronizing procurement and production plans on anticipated tea quality and yield.

Deployment may involve:

Edge Devices (NVIDIA Jetson, Raspberry Pi) in tea processing factories for in-field inference.
Cloud Platforms (AWS, IBM Cloud, Azure) for batch model updates, batched data processing, and analytics.
IoT Integration with field sensors for real-time monitoring.

Cost Implications

Category Cost Component Notes

Hardware Cameras, drones, IoT sensors, edge devices Initial investment; cost increases with size.

Software & Model

Cloud hosting, API services, software licenses
Can be reduced with open-source tools.

Training & Maintenance

Model retraining, sensor calibration
Seasonal or annual expense.

Labour Efficiency

Reduced manual grading, monitoring labor
Easy savings; faster processing eliminates wastefulness.

Return on Investment

Increased yield, better quality grading, lower losses
ROI can be achieved within 1–2 harvesting cycles.

Example: Automatic grading system for tea leaves can be priced at $15,000–$25,000 for estates with medium size, but with ~30% lower labor cost and better compliance for export quality to earn more revenue.

Scalability

In the Tea Industry:

Scale from single estate deployments to multi-estate networks that share cloud infrastructure.
Use centralized dashboards to see in real-time by region.

Technological Scaling

Start with vision-based grading and thereafter include predictive analytics for yield and disease outbreaks.

Integration with ERP systems for end-to-end visibility across the supply chain.

Limitations

High initial capital cost – Unaffordable for smallholders unless made accessible through cooperatives or government programs.
Connectivity Issues – Internet is intermittent in remote plantations; edge computing with periodic syncing required.
Model Drift – A change in tea type, weather, or pest infestation can reduce accuracy with time; requires retraining.
Data Availability – A lack of proper historical datasets in certain regions might limit model performance.
User Training – Field officers must be trained on using equipment and interpreting results.

Automation Potential

Automated Sorting Lines – Computer vision-equipped conveyors to sort tea grades in real-time.
Drone-Based Photoshots of Crop Health – AI-powered drones to scan plantations independently and identify problem areas.
Auto Irrigation Control – Integrate yield forecast with smart irrigation for optimal use of water.
Combined Quality and Packaging – Output grading could be directly interfaced with packing machines for consistency.

Photoshots to Future Applications Beyond Tea

The very same AI + IoT model could be extrapolated to:

Coffee – Bean grading, defects, and estimate of yield.
Cocoa – Detection of cocoa pod maturity and disease control.
Spices – Turmeric or cardamom quality categorization.
Rice & Cereals – Sorting for automatic quality and moisture determination of grains.

Fruit Orchards – Identification of ripeness and auto-harvesting scheduling.

The cross-crop suitability makes this model suitable to agro-tech solution providers, who can serve multiple industries with minimal modification.

Industrial Integration Strategic Roadmap

Pilot Implementation – Execute on one estate to prove the concept.
Stakeholder Training – Share operational knowledge with managers and employees.
Integration with Existing Infrastructure – Integrate grading, storage, and ERP facilities.
Periodic Model Retraining – Incorporate seasonal and regional trends.
Scaling to Multiple Estates & Crops – Geographical scaling and across product lines.

CONCLUSION

This project emphasized the effectiveness of using computer vision technology to the food sector, and more to tea leaves classification. Through the incorporation of powerful image preprocessing, PSO-based feature extraction, and Remora Optimization-based feature subset selection, we were able to boost the quality of data set and classify to the optimal level. Using ResNet-50 for image classification was overall successful with good precision in the identification of various tea leaves by appearance. The results are evidence of the capabilities of computer vision systems in sorting and quality checking foods and the feasibility to be applied to other uses of foods. The utilization of these methods en masse makes the processes more effective and also acts as a steppingstone to more advanced and automated food processing and analysis systems.

REFERENCES

Abdolmaleki, M., Consens, M., & Esmaeili, K. (2022). Ore-Waste Discrimination Using Supervised and Unsupervised Classification of Hyperspectral Images. *Remote Sensing*, *14*, 6386.

Abdon, A. (2015). Numerical analysis of time fractional three-dimensional diffusion equation. *Thermal Science*, *19*, 7–12.

Akbarzadeh, S., Paap, A., Ahderom, S., Apopei, B., & Alameh, K. (2018). Plant discrimination by Support Vector Machine classifier based on spectral reflectance. *Computers and Electronics in Agriculture*, *148*, 250–258.

Amneh, A. A. (2019). Application of Moving Average Filter for the Quantitative Analysis of the NIR Spectra. *Journal of Analytical Chemistry*, *74*, 686–692.

Atangana, A., & Goufo, E. F. D. (2014). Computational analysis of the model describing HIV infection of CD4(+)T cells. *BioMed Research International*, •••, 2014. PMID: 25136605

Bøhn, S. K., Croft, K. D., Burrows, S., Puddey, I. B., Mulder, T. P. J., Fuchs, D., Woodman, R. J., & Hodgson, J. M. (2014). Effects of black tea on body composition and metabolic outcomes related to cardiovascular disease risk: A randomized controlled trial. *Food & Function*, *5*, 1613–1620. PMID: 24889137

Borah, S., Hines, E. L., & Bhuyan, M. (2007). Wavelet transform-based image texture analysis for size estimation applied to the sorting of tea granules. *Journal of Food Engineering*, *79*, 629–639.

Cattani, C. (2010). Harmonic wavelet approximation of random, fractal, and high frequency signals. *Telecommunication Systems*, *43*, 207–217.

Cattani, C. (2012). Fractional calculus and Shannon wavelet. *Mathematical Problems in Engineering*, •••, 2012.

Cattani, C., Pierro, G., & Altieri, G. (2012). Entropy and multifractality for the myeloma multiple TET 2 gene. *Mathematical Problems in Engineering*, •••, 2012.

Chen, Q., Zhao, J., & Cai, J. (2008). Identification of tea varieties using computer vision. *Transactions of the ASABE, 51*, 623–628.

Chen, Q. S., Liu, A. P., Zhao, J. W., & Ouyang, Q. (2013). Classification of tea category using a portable electronic nose based on an odor imaging sensor array. *Journal of Pharmaceutical and Biomedical Analysis, 84*, 77–83. PMID: 23810847

Chen, Q. S., Zhao, J. W., Fang, C. H., & Wang, D. M. (2007). Feasibility study on identification of green, black and oolong teas using near-infrared reflectance spectroscopy based on support vector machine (SVM). *Spectrochimica Acta. Part A: Molecular and Biomolecular Spectroscopy*, *66*, 568–574. PMID: 16859975

De Almeida, V. E., da Costa, G. B., Fernandes, D. D. S., Diniz, P. H. G. D., Brandão, D., de Medeiros, A. C. D., & Véras, G. (2014). Using color histograms and SPA-LDA to classify bacteria. *Analytical and Bioanalytical Chemistry*, *406*, 5989–5995. PMID: 25023972

Fang, L. T., Wu, L. N., & Zhang, Y. D. (2015). A novel demodulation system based on continuous wavelet transform. *Mathematical Problems in Engineering*, •••, 2015.

Gill, G. S., Kumar, A., & Agarwal, R. (2011). Monitoring and grading of tea by computer vision — A review. *Journal of Food Engineering*, *106*, 13–19.

Hajiaghaalipour, F., Kanthimathi, M. S., Sanusi, J., & Rajarajeswaran, J. (2015). White tea (Camellia sinensis) inhibits proliferation of the colon cancer cell line, HT-29, activates caspases, and protects DNA of normal cells against oxidative damage. *Food Chemistry*, *169*, 401–410. PMID: 25236244

Herrador, M. A., & González, A. G. (2001). Pattern recognition procedures for differentiation of green, black, and oolong teas according to their metal content from inductively coupled plasma atomic emission spectrometry. *Talanta*, *53*, 1249–1257. PMID: 18968219

Horanni, R., & Engelhardt, U. H. (2013). Determination of amino acids in white, green, black, oolong, pu-erh teas, and tea products. *Journal of Food Composition and Analysis*, *31*, 94–100.

Jian, W., Xian, Y. Z., & Shi, P. D. (2010). Identification and grading of tea using computer vision. *Applied Engineering in Agriculture*, *26*, 639–645.

Laddi, A., Sharma, S., Kumar, A., & Kapur, P. (2013). Classification of tea grains based upon image texture feature analysis under different illumination conditions. *Journal of Food Engineering*, *115*, 226–231.

Li, X. N., Li, T. P., Gao, X. B., & Cui, Y. (2023). Shielding performance evaluation of smoke screens and revision of the smoke screen mass extinction coefficient. *Bingqi Zhuangbei Gongcheng Xuebao*, *44*, 187–194.

Lim, H. J., Shim, S. B., Jee, S. W., Lee, S. H., Lim, C. J., Hong, J. T., Sheen, Y. Y., & Hwang, D. Y. (2013). Green tea catechin leads to global improvement among Alzheimer's disease-related phenotypes in NSE/hAPP-C105 Tg mice. *The Journal of Nutritional Biochemistry*, *24*, 1302–1313. PMID: 23333093

Liu, N., Liang, Y. Z., Bin, J., Zhang, Z. M., Huang, J. H., Shu, R. X., & Yang, K. (2014). Classification of green and black teas by PCA and SVM analysis of cyclic voltammetric signals from metallic oxide-modified electrode. *Food Analytical Methods*, *7*, 472–480.

Long, W. J., Lei, G. H., Guan, Y. T., Chen, H. Y., Hu, Z. K., She, Y. B., & Fu, H. Y. (2023). Classification of Chinese traditional cereal vinegars and antioxidant property predication by fluorescence spectroscopy. [PubMed]. *Food Chemistry*, *424*, 136406. PMID: 37216781

Luo, X. L., Sun, C. J., He, Y., Zhu, F. L., & Li, X. L. (2023). Cross-cultivar prediction of quality indicators of tea based on VIS-NIR hyperspectral imaging. *Industrial Crops and Products*, *202*, 117009.

Prey, L., Hanemann, A., Ramgraber, L., Seidl-Schulz, J., & Noack, P. O. (2022). UAV-Based Estimation of Grain Yield for Plant Breeding: Applied Strategies for Optimizing the Use of Sensors, Vegetation Indices, Growth Stages, and Machine Learning Algorithms. *Remote Sensing*, *14*, 6345.

Qi, H., & Li, S. X. (2014). Dose-response meta-analysis on coffee, tea, and caffeine consumption with risk of Parkinson's disease. *Geriatrics & Gerontology International, 14*, 430–439. PMID: 23879665

Qian, L. Y., Wu, D. C., Liu, D., Zhou, X. J., Wei, W., Zhong, L. J., Wang, W. J., Wang, Y. J., & Gong, W. (2021). Analysis and Design of Hyperspectral lmaging LiDAR Scanning Mirror. *Acta Optica Sinica, 41*, 232–238.

Sironi, E., Colombo, L., Lompo, A., Messa, M., Bonanomi, M., Regonesi, M. E., Salmona, M., & Airoldi, C. (2014). Natural compounds against neurodegenerative diseases: Molecular characterization of the interaction of catechins from green tea with Aβ1-42, PrP106–126, and Ataxin-3 oligomers. *Chemistry (Weinheim an der Bergstrasse, Germany), 20*, 13793–13800. PMID: 25179684

Tai, Y.-H., Chou, L.-S., & Chiu, H.-L. (2011). Gap-type a-Si TFTs for front light sensing application. *Journal of Display Technology, 7*, 679–683.

Wang, F., Zhao, C. J., Xu, B., Xu, Z., Li, Z. H., Yang, H. B., Duan, D. D., & Yang, G. J. (2020). Development of a portable detection device for the quality of fresh tea leaves using spectral technology. *Nongye Gongcheng Xuebao (Beijing), 36*, 273–280.

Wang, L. F., Zhang, X. W., Liu, J., Shen, L., & Li, Z. Q. (2014). Tea consumption and lung cancer risk: A meta-analysis of case-control and cohort studies. *Nutrition (Burbank, Los Angeles County, Calif.), 30*, 1122–1127. PMID: 25194612

Wang, N., Xing, K. W., Zhang, W., Jiang, L. Z., Elfalleh, W., Cheng, J. J., & Yu, D. Y. (2023). Combining multi–spectroscopy analysis and interfacial properties to research the effect of ultrasonic treatment on soybean protein isolate–tannic acid complexes. *Food Hydrocolloids, 145*, 109136.

Wang, Y. W., Ren, Y. Q., Kang, S. Y., Yin, C. B., Shi, Y., & Men, H. (2023). Identification of tea quality at different picking periods: A hyperspectral system coupled with a multibranch kernel attention network. *Food Chemistry, 433*, 137307. PMID: 37683489

Wu, D., Chen, X. J., & He, Y. (2009). Application of multispectral image texture to discriminating tea categories based on DCT and LS-SVM. *Guangpuxue Yu Guangpu Fenxi*, *29*, 1382–1385. PMID: 19650495

Yang, C. S., & Landau, J. M. (2000). Effects of tea consumption on nutrition and health. *The Journal of Nutrition*, *130*, 2409–2412. PMID: 11015465

Yiannakopoulou, C. (2014). Green tea catechins: Proposed mechanisms of action in breast cancer focusing on the interplay between survival and apoptosis. *Anti-cancer Agents in Medicinal Chemistry*, *14*, 290–295. PMID: 24069935

Zhang, D., Liang, P., Ye, J., Xia, J., Zhou, Y., Huang, J., Ni, D., Tang, L., Jin, S., & Yu, Z. (2019). Detection of systemic pesticide residues in tea products at trace level based on SERS and verified by GC–MS. *Analytical and Bioanalytical Chemistry*, *411*, 7187–7196. PMID: 31620825

Zhang, S., Wang, S., & Wu, L. (2010). A novel method for magnetic resonance brain image classification based on adaptive chaotic PSO. *Progress in Electromagnetics Research*, *109*, 325–343.

Zhang, Y., Zhou, J. H., Wang, S. M., Wang, Y. Y., Zhang, Y. H., Zhao, S., Liu, S. Y., & Yang, J. (2023). Identification of Xinhui Citri Reticulatae Pericarpium of Different AgingYears Based on Visible Near Infrared Hyperspectral lmaging. *Guangpuxue Yu Guangpu Fenxi*, *43*, 3286–3292.

Zhang, Y. D., Wang, S. H., Ji, G. L., & Dong, Z. C. (2015). Exponential wavelet iterative shrinkage thresholding algorithm with random shift for compressed sensing magnetic resonance imaging. *IEEJ Transactions on Electrical and Electronic Engineering*, *10*, 116–117.

Zhang, Y. D., Wang, S. H., Ji, G. L., & Phillips, P. (2014). Fruit classification using computer vision and feedforward neural network. *Journal of Food Engineering*, *143*, 167–177.

Zhang, Y. D., Wang, S. H., Phillips, P., Dong, Z. C., Ji, G. L., & Yang, J. Q. (2015). Detection of Alzheimer's disease and mild cognitive impairment based on structural volumetric MR images using 3D-DWT and WTA-KSVM trained by PSOTVAC. *Biomedical Signal Processing and Control*, *21*, 58–73.

Zhao, J. W., Chen, Q. S., Huang, X. Y., & Fang, C. H. (2006). Qualitative identification of tea categories by near infrared spectroscopy and support vector machine. *Journal of Pharmaceutical and Biomedical Analysis*, *41*, 1198–1204. PMID: 16621404

Compilation of References

Abdolmaleki, M., Consens, M., & Esmaeili, K. (2022). Ore-Waste Discrimination Using Supervised and Unsupervised Classification of Hyperspectral Images. *Remote Sensing*, *14*, 6386.

Abdon, A. (2015). Numerical analysis of time fractional three-dimensional diffusion equation. *Thermal Science*, *19*, 7–12.

Ahmadi, A. (2023). The AI food revolution: Reshaping food sciences through artificial intelligence. *International Journal of BioLife Sciences*, *2*(1), 62–71. DOI: 10.22034/jbs.2023.178662

Aipoh, G., Osunkanmbi, A. A., & Egbemhenghe, J. (2025). Artificial intelligence in logistics and distribution: The function of AI in dynamic route planning for transportation, including self-driving trucks and drone delivery systems. *World Journal of Advanced Research and Reviews*, *25*(2), 155–167. DOI: 10.30574/wjarr.2025.25.2.0214

Akbarzadeh, S., Paap, A., Ahderom, S., Apopei, B., & Alameh, K. (2018). Plant discrimination by Support Vector Machine classifier based on spectral reflectance. *Computers and Electronics in Agriculture*, *148*, 250–258.

Alhammad, S. M. (2025). *Deep learning and explainable AI for classification of plant diseases*. Front. AI.

Amneh, A. A. (2019). Application of Moving Average Filter for the Quantitative Analysis of the NIR Spectra. *Journal of Analytical Chemistry*, *74*, 686–692.

Amuthadevi, C., Banu, E. A., Kumar, S. S., Karthick, S., Kistan, A., & Sudhakar, M. (2025). Aeroponics Techniques for Improved Farming Using Artificial and Deep Learning techniques. In *Utilizing Aeroponics* Techniques for Improved Farming (pp. 81-118). IGI Global Scientific Publishing.

Anusha, S., Ezhilvandan, M., Gayathri, S., Sivaraman, V., & TP, A. (2024). Emerging trends in edge AI for Industry 4.0 and 5.0: Technologies, applications, and challenges. AI Innovations in Service and Tourism Marketing, 211-228.

Anusha, S., Yamini, B., Udhayashankar, S., Priya, V., Ezhilvendan, M., Vinoth, R., & Subramanian, R. S. (2025). AI and Applied Microbiology in Sustainable Waste Recycling and the Circular Economy. In *AI Technologies for Enhancing Recycling Processes* (pp. 83–100). IGI Global Scientific Publishing.

Ascherio, A., Zhang, S. M., Hernán, M. A., Kawachi, I., Colditz, G. A., Speizer, F. E., & Willett, W. C. (2001). Prospective study of caffeine consumption and risk of Parkinson's disease in men and women. *Annals of Neurology*, *50*, 56–63. PMID: 11456310

Atangana, A., & Goufo, E. F. D. (2014). Computational analysis of the model describing HIV infection of CD4(+)T cells. *BioMed Research International*, •••, 2014. PMID: 25136605

Babatunde, R. S. (2024). *A novel smartphone application for early detection of plant diseases. (app/prototype & field study)*. Academic Press.

Balakrishnan, C., Anish, T. P., Nithya, T., Sheena, B. G., Maheswari, B., & Subramanian, R. S. (2024). Revolutionizing Trust Manufacturing Integrating Real-Time Data and AI Innovations for Enhanced Efficiency. In Industry Applications of Thrust Manufacturing: Convergence with Real-Time Data and AI (pp. 74-97). IGI Global.

Batista, L. R., Chalfoun de Souza, S. M., Silva e Batista, C. F., & Schwan, R. F. (2015). Coffee: Types and production. In *Encyclopedia of Food and Health* (pp. 121-125). Elsevier Ltd.

Bhatt, S. (2024). A Comprehensive Analysis of Literature and Strategy for Eliminating Food Waste in Food Service Industry. *Promoting Sustainable Gastronomy Tourism and Community Development*, 201-213.

Bøhn, S. K., Croft, K. D., Burrows, S., Puddey, I. B., Mulder, T. P. J., Fuchs, D., Woodman, R. J., & Hodgson, J. M. (2014). Effects of black tea on body composition and metabolic outcomes related to cardiovascular disease risk: A randomized controlled trial. *Food & Function*, *5*, 1613–1620. PMID: 24889137

Bolka, M., & Emire, S. (2020). Effects of coffee roasting technologies on cup quality and bioactive compounds of specialty coffee beans. *Food Science & Nutrition*, *8*, 6120–6130. PMID: 33282263

Boopathi, S. (2024). Advancements in machine learning and AI for intelligent systems in drone applications for smart city developments. In *Futuristic e-Governance Security With Deep Learning Applications* (pp. 15–45). IGI Global Scientific Publishing.

Borah, S., Hines, E. L., & Bhuyan, M. (2007). Wavelet transform-based image texture analysis for size estimation applied to the sorting of tea granules. *Journal of Food Engineering*, *79*, 629–639.

Brahimi, M., Boukhalfa, K., & Moussaoui, A. (2017). Deep learning for tomato diseases: Classification and symptoms visualization. *Applied Artificial Intelligence*, *31*(4), 299–315.

Camandola, S., Plick, N., & Mattson, M. P. (2019). Impact of coffee and cacao purine metabolites on neuroplasticity and neurodegenerative disease. *Neurochemical Research*, *44*, 214–227. PMID: 29417473

Carmona-Torres, C., Dev, K., Mhlanga, D., & Aït-Kaddour, A. (2024). From Food Industry 4.0 to Food Industry 5.0: Identifying technological enablers and potential future applications in the food sector. *Comprehensive Reviews in Food Science and Food Safety*, *23*(6), e370040. DOI: 10.1111/1541-4337.70040 PMID: 39437193

Cattani, C. (2010). Harmonic wavelet approximation of random, fractal, and high frequency signals. *Telecommunication Systems*, *43*, 207–217.

Cattani, C. (2012). Fractional calculus and Shannon wavelet. *Mathematical Problems in Engineering*, ●●●, 2012.

Cattani, C., Pierro, G., & Altieri, G. (2012). Entropy and multifractality for the myeloma multiple TET 2 gene. *Mathematical Problems in Engineering*, ●●●, 2012.

Chamara, R. M. S. R., Senevirathne, N. N., De Silva, S. H. N. P., Ariyaratne, W. M. T. P., & Marambe, B. (2020). Role of artificial intelligence in achieving global food security: A promising technology for the future. *Sri Lanka Journal of Food and Agriculture*, *6*(2), 43–70. DOI: 10.4038/sljfa.v6i2.88

Chen, Q. S., Liu, A. P., Zhao, J. W., & Ouyang, Q. (2013). Classification of tea category using a portable electronic nose based on an odor imaging sensor array. *Journal of Pharmaceutical and Biomedical Analysis*, *84*, 77–83. PMID: 23810847

Chen, Q. S., Zhao, J. W., Fang, C. H., & Wang, D. M. (2007). Feasibility study on identification of green, black and oolong teas using near-infrared reflectance spectroscopy based on support vector machine (SVM). *Spectrochimica Acta. Part A: Molecular and Biomolecular Spectroscopy*, *66*, 568–574. PMID: 16859975

Chen, Q., Zhao, J., & Cai, J. (2008). Identification of tea varieties using computer vision. *Transactions of the ASABE*, *51*, 623–628.

Chhikara, H., Chhikara, S., & Gupta, L. (2025). Predictive Analytics in Finance: Leveraging AI and Machine Learning for Investment Strategies. In Utilizing AI and Machine Learning in Financial Analysis (pp. 325-336). IGI Global Scientific Publishing.

Chinnasamy, P., Sathya, K. S., Jebamani, B. J. A., Nithyasri, A., & Fowjiya, S. (2023). Deep learning: algorithms, techniques, and applications—a systematic survey. In *Deep learning research applications for natural language processing* (pp. 1–17). IGI global.

Creswell, J. W., & Poth, C. N. (2018). *Qualitative Inquiry and Research Design: Choosing Among Five Approaches* (4th ed.). SAGE Publications.

Das, S. (2017). *CNN architectures: LeNet, AlexNet, VGG, GoogLeNet, ResNet, and more*. Medium.

De Almeida, V. E., da Costa, G. B., Fernandes, D. D. S., Diniz, P. H. G. D., Brandão, D., de Medeiros, A. C. D., & Véras, G. (2014). Using color histograms and SPA-LDA to classify bacteria. *Analytical and Bioanalytical Chemistry*, *406*, 5989–5995. PMID: 25023972

Dhinakaran, D., Kumar, N. J., & Ponnuviji, N. P. (2025). Safeguarding confidentiality and privacy in cloud-enabled healthcare systems with spectrasafe encryption and dynamic k-anonymity algorithm. *Expert Systems with Applications*, *279*, 127584.

Ding, H., Tian, J., Yu, W., Wilson, D. I., Young, B. R., Cui, X., Xin, X., Wang, Z., & Li, W. (2023). The Application of Artificial Intelligence and Big Data in the Food Industry. *Foods*, *12*(4511). PMID: 38137314

EASA. (2024). Easy Access Rules for Unmanned Aircraft Systems. https://www.easa.europa.eu/sites/default/files/dfu/D0593E_2024-07-10_06.26.37_EAR-for-Unmanned-Aircraft-Systems.pdf

Fang, L. T., Wu, L. N., & Zhang, Y. D. (2015). A novel demodulation system based on continuous wavelet transform. *Mathematical Problems in Engineering*, •••, 2015.

Federal Aviation Administration (FAA). (2023). *Part 107 Small Unmanned Aircraft Systems Regulations*. https://www.faa.gov/newsroom/small-unmanned-aircraft-systems-uas-regulations-part-107

Ferentinos, K. P. (2018). Deep learning models for plant disease detection and diagnosis. *Computers and Electronics in Agriculture*, *145*, 311–318.

Ferreira, L. C. (2024). Hyperspectral imaging for detection of plant pathogens — Review and advances. *Front. Plant Sci.*

Food and Agriculture Organization. (2021). The State of Food Security and Nutrition in the World. FAO.

Ganesan, I., Ponnuviji, N. P., Kumar, A. S., Nithya, M., Jambulingam, U., & Lalitha, S. D. (2024). Real-Time Event Detection and Predictive Analytics Using IoT and Deep Learning. In Industry Applications of Thrust Manufacturing: Convergence with Real-Time Data and AI (pp. 1-41). IGI Global Scientific Publishing.

Gill, G. S., Kumar, A., & Agarwal, R. (2011). Monitoring and grading of tea by computer vision — A review. *Journal of Food Engineering*, *106*, 13–19.

Gracious, L. A., Jasmine, R. M., Pooja, E., Anish, T. P., Johncy, G., & Subramanian, R. S. (2023, October). Machine Learning and Deep Learning Transforming Healthcare: An Extensive Exploration of Applications, Algorithms, and Prospects. In *2023 4th IEEE Global Conference for Advancement in Technology (GCAT)* (pp. 1-6). IEEE.

Grewal, D., Guha, A., & Noble, S. M.. (2024). The food production–consumption chain: Fighting food insecurity, loss, and waste with technology. *Journal of the Academy of Marketing Science*, *52*, 1412–1430. DOI: 10.1007/s11747-024-01040-x

Grobbelaar, W., Verma, A., & Shukla, V. K. (2021). Analyzing human robotic interaction in the food industry. *Journal of Physics: Conference Series*, *1714*(1), 012032. DOI: 10.1088/1742-6596/1714/1/012032

Hajiaghaalipour, F., Kanthimathi, M. S., Sanusi, J., & Rajarajeswaran, J. (2015). White tea (Camellia sinensis) inhibits proliferation of the colon cancer cell line, HT-29, activates caspases, and protects DNA of normal cells against oxidative damage. *Food Chemistry*, *169*, 401–410. PMID: 25236244

Hassanien, A. E., & Darwish, A. (Eds.). (2020). *Machine Learning and Artificial Intelligence in Agricultural Research*. Springer.

Herrador, M. A., & González, A. G. (2001). Pattern recognition procedures for differentiation of green, black, and oolong teas according to their metal content from inductively coupled plasma atomic emission spectrometry. *Talanta*, *53*, 1249–1257. PMID: 18968219

Hiremath, V. R., & Ingalagi, S. (2025). Optimizing Digital Marketing Strategies With Machine Learning and Deep Learning Innovations. In Impact of Digital Transformation on Business Growth and Performance (pp. 175-204). IGI Global Scientific Publishing.

Horanni, R., & Engelhardt, U. H. (2013). Determination of amino acids in white, green, black, oolong, pu-erh teas, and tea products. *Journal of Food Composition and Analysis*, *31*, 94–100.

Huang, N. F., Chou, D. L., Lee, C. A., Wu, F. P., Chuang, A. C., Chen, Y. H., & Tsai, Y. C. (2020). Smart agriculture: Real-time classification of green coffee beans by using a convolutional neural network. *IET Smart Cities*, *2*(167–172).

Hu, G., Bidel, S., Jousilahti, P., Antikainen, R., & Tuomilehto, J. (2007). Coffee and tea consumption and the risk of Parkinson's disease. *Movement Disorders*, *22*, 2242–2248. PMID: 17712848

Hwa, L. S., & Chuan, L. T. (2024). A brief review of artificial intelligence robotics in the food industry. *Procedia Computer Science*, *232*, 1694–1700. DOI: 10.1016/j.procs.2024.01.167

Hwa, L. S., & Te Chuan, L. (2024). A brief review of artificial intelligence robotic in food industry. *Procedia Computer Science*, *232*, 1694–1700.

Iqab, M. (2024). *Harnessing drones for faster, cheaper, and greener logistic solutions in challenging environments* (Bachelor's thesis). LAB University of Applied Sciences. https://urn.fi/URN:NBN:fi:amk-202405039113

Jahani, H., Khosravi, Y., Kargar, B., Ong, K.-L., & Arisian, S. (2024). Exploring the role of drones and UAVs in logistics and supply chain management: A novel text-based literature review. *International Journal of Production Research*. Advance online publication. DOI: 10.1080/00207543.2024.2373425

Jebamalar, B., Sudha, K., Maheswari, B., Nalini, M., Dinesh, M. G., & Subramanian, R. S. (2023, May). Collaborative filtering for bargaining and customer feedback in e-commerce. In 2023 7th International Conference on Intelligent Computing and Control Systems (ICICCS) (pp. 1804-1808). IEEE.

Jian, W., Xian, Y. Z., & Shi, P. D. (2010). Identification and grading of tea using computer vision. *Applied Engineering in Agriculture*, *26*, 639–645.

Kachouei, M. A., Kaushik, A., & Ali, M. A. (2023). Internet of Things-enabled food and plant sensors to empower sustainability. *Advanced Intelligent Systems*, *5*(12), 2300321. DOI: 10.1002/aisy.202300321

Kapoor, A., Vo, D.-V. N., & Rangasamy, G. (2023). Digitalization of the agro-food sector for achieving sustainable development goals: A review. *Sustainable Food Technology*, *1*, 783–802. DOI: 10.1039/d3fb00124e

Kashyap, R. (2019). Sensation of deep learning in image processing applications. In *Handbook of Research on Deep Learning Innovations and Trends* (pp. 72–96). IGI Global.

Keogh, J. G., Dube, L., Rejeb, A., Hand, K. J., Khan, N., & Dean, K. (2020). The future food chain: digitization as an enabler of Society 5.0. In *Building the Future of Food Safety Technology* (pp. 11–38). Elsevier., DOI: 10.1016/b978-0-12-818956-6.00002-6

Khang, A. (Ed.). (2023). *Handbook of research on AI-equipped IoT applications in high-tech agriculture*. IGI Global.

Khayyam, M., Ihtisham, M., & Asongu, S. A. (2021). Potential role of technology innovation in transformation of sustainable food systems. *Revista de Agricultura (Piracicaba)*, *11*(984). Advance online publication. DOI: 10.3390/agriculture11100984

Kitonsa, H., & Kruglikov, S. V. (2018). Significance of drone technology for achievement of the United Nations sustainable development goals. *R-economy*, *4*(3), 115–120. DOI: 10.15826/recon.2018.4.3.016

Kouadio, L. (2023). A review on UAV-based applications for plant disease detection. *Remote Sensing*.

Król, K., Gantner, M., Tatarak, A., & Hallmann, E. (2020). The content of polyphenols in coffee beans as roasting, origin and storage effect. *European Food Research and Technology*, *246*, 33–39.

Kumar, A., Bhatia, A., Kashyap, A., & Kumar, M. (2023). LSTM network: a deep learning approach and applications. In *Advanced Applications of NLP and Deep Learning in Social Media Data* (pp. 130–150). IGI Global.

Kumar, A., Singh, R., & Kumar, S. (2020). Application of Artificial Intelligence in Food Quality Control: A Review. *Journal of Food Science and Technology*, *57*(10), 3771–3780.

Kumar, R., & Dwivedi, R. K. (Eds.). (2022). *Advances in Deep Learning Applications for Smart Cities*. IGI Global.

Laddi, A., Sharma, S., Kumar, A., & Kapur, P. (2013). Classification of tea grains based upon image texture feature analysis under different illumination conditions. *Journal of Food Engineering*, *115*, 226–231.

Liczbiński, P., & Bukowska, B. (2022). Tea and coffee polyphenols and their biological properties based on the latest in vitro investigations. *Industrial Crops and Products*, *175*, 114265. PMID: 34815622

Lim, H. J., Shim, S. B., Jee, S. W., Lee, S. H., Lim, C. J., Hong, J. T., Sheen, Y. Y., & Hwang, D. Y. (2013). Green tea catechin leads to global improvement among Alzheimer's disease-related phenotypes in NSE/hAPP-C105 Tg mice. *The Journal of Nutritional Biochemistry*, *24*, 1302–1313. PMID: 23333093

Liu, N., Liang, Y. Z., Bin, J., Zhang, Z. M., Huang, J. H., Shu, R. X., & Yang, K. (2014). Classification of green and black teas by PCA and SVM analysis of cyclic voltammetric signals from metallic oxide-modified electrode. *Food Analytical Methods*, *7*, 472–480.

Li, X. N., Li, T. P., Gao, X. B., & Cui, Y. (2023). Shielding performance evaluation of smoke screens and revision of the smoke screen mass extinction coefficient. *Bingqi Zhuangbei Gongcheng Xuebao*, *44*, 187–194.

Long, W. J., Lei, G. H., Guan, Y. T., Chen, H. Y., Hu, Z. K., She, Y. B., & Fu, H. Y. (2023). Classification of Chinese traditional cereal vinegars and antioxidant property predication by fluorescence spectroscopy. [PubMed]. *Food Chemistry*, *424*, 136406. PMID: 37216781

Luo, X. L., Sun, C. J., He, Y., Zhu, F. L., & Li, X. L. (2023). Cross-cultivar prediction of quality indicators of tea based on VIS-NIR hyperspectral imaging. *Industrial Crops and Products*, *202*, 117009.

Mahesh, B. (2020). Machine learning algorithms-a review. *International Journal of Science and Research (IJSR).[Internet]*, *9*(1), 381-386.

Maheswari, B., & Nagarajan, R. (2024). EduChatbot: Implementing educational Chatbot for assisting the teaching-learning process by NLP-based hybrid heuristic adopted deep learning framework. *Kybernetes*.

Miranda, J., Ponce, P., Molina, A., & Wright, P. (2019). Sensing, smart and sustainable technologies for Agri-Food 4.0. *Computers in Industry*, *108*, 21–36. DOI: 10.1016/j.compind.2019.02.002

Mitiku, H., Kim, T. Y., Kang, H., Apostolidis, E., Lee, J. Y., & Kwon, Y. I. (2022). Selected coffee (*Coffea arabica* L.) extracts inhibit intestinal α-glucosidases activities in-vitro and postprandial hyperglycemia in SD rats. *BMC Complementary Medicine and Therapies*, *22*, 249. PMID: 36151573

Mohanty, S. P., Hughes, D. P., & Salathé, M. (2016). Using deep learning for image-based plant disease detection. *Frontiers in Plant Science*, *7*, 1419. PMID: 27713752

Nguyen, T. M., Quach, S., & Thaichon, P. (2022). The effect of AI quality on customer experience and brand relationship. *Journal of Consumer Behaviour*, *21*(3), 481–493.

Nithya, T., Kumar, V. N., & Deepa, S., CM, V., & Subramanian, R. S. (2023, August). A comprehensive survey of machine learning: Advancements, applications, and challenges. In *2023 Second International Conference on Augmented Intelligence and Sustainable Systems (ICAISS)* (pp. 354-361). IEEE.

Oluwaseun, F. M., Bolarinwa, E. J., & Nnanna, O. C. (2024). Smart Farming [IJAEM]. *Ai Technology for Farm Monitoring and Security in Nigeria International Journal of Advances in Engineering and Management*, *6*(08), 771–781.

Pacal, I. (2024). *A systematic review of deep learning techniques for plant disease detection (2020–2024).* Academic Press.

Pajila, P. B., Sudha, K., Selvi, D. K., Kumar, V. N., Gayathri, S., & Subramanian, R. S. (2023, July). A survey on natural language processing and its applications. In *2023 4th International Conference on Electronics and Sustainable Communication Systems (ICESC)* (pp. 996-1001). IEEE.

Patel, M., Sinha, A., & Desai, R. (2022). Weather-based forecasting in agricultural logistics. *Applied Artificial Intelligence*, *36*(1), 45–63.

Ponnuviji, N. P., Nirmala, G., Kokila, M. S., & Priyadharshini, S. I. (2025). Optimizing Image Retrieval in Cloud Servers with TN-AGW: A Secure and Efficient Approach. Journal of The Institution of Engineers (India): Series B, 106(2), 459-473.

Prabha, D., Subramanian, R. S., Dinesh, M. G., & Girija, P. (2024). Sustainable Farming Through AI-Enabled Precision Agriculture. In Artificial Intelligence for Precision Agriculture (pp. 159-182). Auerbach Publications.

Prabha, D., Maheswari, B., & Aswini, J. (2022). Wrapper-Naive Bayes approach to perform efficient customer behavior prediction. In Innovative Data Communication Technologies and Application [Singapore: Springer Nature Singapore.]. *Proceedings of ICIDCA, 2021*, 17–31.

Prabha, D., Subramanian, R. S., Dinesh, M. G., & Girija, P. (2024). Sustainable Farming Through AI-Enabled Precision Agriculture. In *Artificial Intelligence for Precision Agriculture* (pp. 159–182). Auerbach Publications.

Prajapati, J. B., Barad, R., Patel, M. B., Saini, K., Prajapati, D., & Engineer, P. (2023). Smart farming ingredients: IoT sensors, software, connectivity, data analytics, robots, drones, GIS-GPS. In *Applying drone technologies and robotics for agricultural sustainability* (pp. 31–49). IGI Global.

Prasad, M. S. C., Dhanalakshmi, M., Mohan, M., Somasundaram, B., Valarmathi, R., & Boopathi, S. (2024). Machine Learning-Integrated Sustainable Engineering and Energy Systems: Innovations at the Nexus. In *Harnessing High-Performance Computing and AI for Environmental Sustainability* (pp. 74-98). IGI Global.

Prey, L., Hanemann, A., Ramgraber, L., Seidl-Schulz, J., & Noack, P. O. (2022). UAV-Based Estimation of Grain Yield for Plant Breeding: Applied Strategies for Optimizing the Use of Sensors, Vegetation Indices, Growth Stages, and Machine Learning Algorithms. *Remote Sensing*, *14*, 6345.

Przybył, K., Samborska, K., Koszela, K., Masewicz, L., & Pawlak, T. (2021). Artificial neural networks in the evaluation of the influence of the type and content of carrier on selected quality parameters of spray dried raspberry powders. *Measurement*, *186*, 110014.

Przybył, K., Wawrzyniak, J., Koszela, K., Adamski, F., & Gawrysiak-Witulska, M. (2020). Application of deep and machine learning using image analysis to detect fungal contamination of rapeseed. *Sensors (Basel)*, *20*(7305). PMID: 33352649

Qian, L. Y., Wu, D. C., Liu, D., Zhou, X. J., Wei, W., Zhong, L. J., Wang, W. J., Wang, Y. J., & Gong, W. (2021). Analysis and Design of Hyperspectral Imaging LiDAR Scanning Mirror. *Acta Optica Sinica*, *41*, 232–238.

Qi, H., & Li, S. (2014). Dose-response meta-analysis on coffee, tea and caffeine consumption with risk of Parkinson's disease. *Geriatrics & Gerontology International*, *14*, 430–439. PMID: 23879665

Qi, H., & Li, S. X. (2014). Dose-response meta-analysis on coffee, tea, and caffeine consumption with risk of Parkinson's disease. *Geriatrics & Gerontology International*, *14*, 430–439. PMID: 23879665

R.S. Prabha, D., Aswini, J., Maheswari, B., & Anita, M. (2021, February). Alleviating NB conditional independence using multi-stage variable selection (MSVS): banking customer dataset application. In Journal of Physics: Conference Series (Vol. 1767, No. 1, p. 012002). IOP Publishing.

R.S. Sudha, K., Ramana, K. V., SivaKumar, S., Nithyanandhan, R., & Nalini, M. (2023, February). Hybrid Variable Selection Approach to Analyse High Dimensional Dataset. In 2023 7th International Conference on Computing Methodologies and Communication (ICCMC) (pp. 1489-1495). IEEE.

Rajendran, T., Ramana, T. V., Nalini, M., Subramanian, R. S., Devi, D. C., & Anuradha, M. (2023, July). Paddy Plant Leaf Disease Classification based on Artificial Intelligence. In 2023 4th International Conference on Electronics and Sustainable Communication Systems (ICESC) (pp. 1002-1009). IEEE.

Rashed, M., Fakhry, S., & Satour, R. (2025). Revolutionizing Food Quality With Machine Vision and Machine Learning Techniques. In *Food in the Metaverse and Web 3.0 Era: Intersecting Food, Technology, and Culture* (pp. 71-124). IGI Global Scientific Publishing.

Reis, C. E. G., Dórea, J. G., & da Costa, T. H. M. (2019). Effects of coffee consumption on glucose metabolism: A systematic review of clinical trials. *Journal of Traditional and Complementary Medicine*, *9*, 184–191. PMID: 31193893

Rusinek, R., Dobrzański, B., Oniszczuk, A., Gawrysiak-Witulska, M., Siger, A., Karami, H., Ptaszyńska, A. A., Żytek, A., Kapela, K., & Gancarz, M. (2022). How to identify roast defects in coffee beans based on the volatile compound profile. *Molecules (Basel, Switzerland)*, *27*(8530). PMID: 36500625

Sambana, B. (2025). An efficient plant disease detection using transfer learning (YOLOv7/8). Sci. Rep. / Nature Communications type article.

Saravanan, K., Prakash, R. B., Balakrishnan, C., Kumar, G. V. P., Subramanian, R. S., & Anita, M. (2023, December). Support Vector Machines: Unveiling the Power and Versatility of SVMs in Modern Machine Learning. In *2023 3rd International Conference on Innovative Mechanisms for Industry Applications (ICIMIA)* (pp. 680-687). IEEE.

Sarker, I. H. (2021). Deep learning: A comprehensive overview on techniques, taxonomy, applications and research directions. *SN Computer Science*, *2*(420). PMID: 34426802

Singh, J. N., Saxena, S., Saini, R., Chauan, M., & Saini, K. (2023, December). The Role of Artificial Intelligence in Information Acquisition. In *2023 5th International Conference on Advances in Computing, Communication Control and Networking (ICAC3N)* (pp. 1153-1157). IEEE.

Sironi, E., Colombo, L., Lompo, A., Messa, M., Bonanomi, M., Regonesi, M. E., Salmona, M., & Airoldi, C. (2014). Natural compounds against neurodegenerative diseases: Molecular characterization of the interaction of catechins from green tea with Aβ1-42, PrP106–126, and Ataxin-3 oligomers. *Chemistry (Weinheim an der Bergstrasse, Germany)*, *20*, 13793–13800. PMID: 25179684

Siva Subramanian, R., & Prabha, D. (2021). Optimizıng Naive Bayes probability estimation in customer analysis using hybrid variable selection. In Computer Networks and Inventive Communication Technologies: Proceedings of Third ICCNCT 2020 (pp. 595-612). Springer Singapore.

Sladojevic, S., Arsenovic, M., Anderla, A., Culibrk, D., & Stefanovic, D. (2016). Deep neural networks based recognition of plant diseases by leaf image classification. *Computational Intelligence and Neuroscience*. PMID: 27418923

Sreelakshmi, S., Babu, A. A., Lakshmipriya, C., Gracious, L. A., Nalini, M., & Subramanian, R. S. (2024, October). Enhancing Intrusion Detection Systems with Machine Learning. In *2024 2nd International Conference on Self Sustainable Artificial Intelligence Systems (ICSSAS)* (pp. 557-564). IEEE.

Subramanian, R. S., Yamini, B., Sudha, K., & Sivakumar, S. (2024). Ensemble-based deep learning techniques for customer churn prediction model. *Kybernetes*.

Sudha, K., Balakrishnan, C., Anish, T. P., Nithya, T., Yamini, B., Subramanian, R. S., & Nalini, M. (2024). Data Insight Unveiled: Navigating Critical Approaches and Challenges in Diverse Domains Through Advanced Data Analysis. Critical Approaches to Data Engineering Systems and Analysis, 90-114.

Sudha, K., Ambhika, C., Maheswari, B., Girija, P., & Nalini, M. (2023). AI and IoT applications in medical domain enhancing healthcare through technology integration. In *AI and IoT-Based Technologies for Precision Medicine* (pp. 280–294). IGI Global.

Sudiyono, K. A., Utomo, P., & Severesia, C. (2022). Effect of customer experience and customer value towards customer loyalty and satisfaction on B2B food and beverage sector. *Journal of Business and Management Review*, *3*(9), 627–640.

Suresh Babu, C. V. Koti Reddy, Chandu, G., Kowshik, T., & Rao, L. N. (2024). Trustworthy AI for Optimizing Agriculture: Leveraging IoT Sensory Data. In S. Sharma, A. Prakash, & V. Sugumaran (Eds.), Developments Towards Next Generation Intelligent Systems for Sustainable Development (pp. 124-147). IGI Global. https://doi.org/DOI: 10.4018/979-8-3693-5643-2.ch005

Suresh Babu, C. V., Yadavamuthiah, K., Abirami, S., & Kowsika, S. (2023). Artificial Intelligence in Wastewater Management. In Shikuku, V. (Ed.), *Artificial Intelligence Applications in Water Treatment and Water Resource Management* (pp. 31–45). IGI Global., DOI: 10.4018/978-1-6684-6791-6.ch002

Tai, Y.-H., Chou, L.-S., & Chiu, H.-L. (2011). Gap-type a-Si TFTs for front light sensing application. *Journal of Display Technology*, *7*, 679–683.

Tajik, N., Tajik, M., Mack, I., & Enck, P. (2017). The potential effects of chlorogenic acid, the main phenolic components in coffee, on health: A comprehensive review of the literature. *European Journal of Nutrition*, *56*, 2215–2244. PMID: 28391515

Tomar, P., & Kaur, G. (Eds.). (2021). *Artificial intelligence and IoT-based technologies for sustainable farming and smart agriculture*. IGI global.

Tsai, C. F., & Jioe, I. P. J. (2021). The analysis of chlorogenic acid and caffeine content and its correlation with coffee bean color under different roasting degree and sources of coffee (*Coffea arabica* Typica). *Processes (Basel, Switzerland)*, *9*(2040).

Venkateswaran, N., Vidhya, R., Naik, D. A., Raj, T. M., Munjal, N., & Boopathi, S. (2023). Study on Sentence and Question Formation Using Deep Learning Techniques. In *Digital Natives as a Disruptive Force in Asian Businesses and Societies* (pp. 252–273). IGI Global Scientific Publishing.

Wang, F., Zhao, C. J., Xu, B., Xu, Z., Li, Z. H., Yang, H. B., Duan, D. D., & Yang, G. J. (2020). Development of a portable detection device for the quality of fresh tea leaves using spectral technology. *Nongye Gongcheng Xuebao (Beijing)*, *36*, 273–280.

Wang, L. F., Zhang, X. W., Liu, J., Shen, L., & Li, Z. Q. (2014). Tea consumption and lung cancer risk: A meta-analysis of case-control and cohort studies. *Nutrition (Burbank, Los Angeles County, Calif.)*, *30*, 1122–1127. PMID: 25194612

Wang, N., Xing, K. W., Zhang, W., Jiang, L. Z., Elfalleh, W., Cheng, J. J., & Yu, D. Y. (2023). Combining multi–spectroscopy analysis and interfacial properties to research the effect of ultrasonic treatment on soybean protein isolate–tannic acid complexes. *Food Hydrocolloids*, *145*, 109136.

Wang, P., Tseng, H. W., Chen, T. C., & Hsia, C. H. (2021). Deep convolutional neural network for coffee bean inspection. *Sensors and Materials*, *33*, 2299–2310.

Wang, W., & Siau, K. (2019). Artificial Intelligence, Machine Learning, Automation, Robotics, Future of Work and Future of Humanity: A Review and Research Agenda. *Journal of Database Management*, *30*, 61–79. DOI: 10.4018/JDM.2019010104

Wang, Y. W., Ren, Y. Q., Kang, S. Y., Yin, C. B., Shi, Y., & Men, H. (2023). Identification of tea quality at different picking periods: A hyperspectral system coupled with a multibranch kernel attention network. *Food Chemistry*, *433*, 137307. PMID: 37683489

Wesołowski, P., & Gawałek, J. (2008). Effect of the conditions of cereal instant coffee granulation on the product yield and quality. *Przemysl Chemiczny*, *87*(311–314).

Wu, D., Chen, X. J., & He, Y. (2009). Application of multispectral image texture to discriminating tea categories based on DCT and LS-SVM. *Guangpuxue Yu Guangpu Fenxi*, *29*, 1382–1385. PMID: 19650495

Yamini, B., Prasanna, V., Ambhika, C., Anuradha, M., Maheswari, B., Subramanian, S. R., & Nalini, M. (2023). A comprehensive survey of deep learning: Advancements, applications, and challenges. *International Journal on Recent and Innovation Trends in Computing and Communication*, *11*, 445–453.

Yamini, B., Ramana, K. V., Nalini, M., Devi, D. C., Maheswari, B., & Siva, S. R. (2024). Customer churn prediction model in enterprises using machine learning. *International Journal of Advanced Technology and Engineering Exploration*, *11*(110), 94.

Yang, C. S., & Landau, J. M. (2000). Effects of tea consumption on nutrition and health. *The Journal of Nutrition*, *130*, 2409–2412. PMID: 11015465

Yesil, A., & Yilmaz, Y. (2013). Review article: Coffee consumption, the metabolic syndrome, and non-alcoholic fatty liver disease. *Alimentary Pharmacology & Therapeutics*, *38*, 1038–1044. PMID: 24024834

Yiannakopoulou, C. (2014). Green tea catechins: Proposed mechanisms of action in breast cancer focusing on the interplay between survival and apoptosis. *Anti-cancer Agents in Medicinal Chemistry*, *14*, 290–295. PMID: 24069935

Yin, R. K. (2018). *Case Study Research and Applications: Design and Methods* (6th ed.). SAGE Publications.

Yu, Y., Cao, Y., Lai, Q., Zhao, Q., Sun, Z., Zhou, S., & Song, D. (2023). Design and operation parameters of vibrating harvester for *Coffea arabica* L. *Agriculture*, *13*(700).

Zhang, D., Liang, P., Ye, J., Xia, J., Zhou, Y., Huang, J., Ni, D., Tang, L., Jin, S., & Yu, Z. (2019). Detection of systemic pesticide residues in tea products at trace level based on SERS and verified by GC–MS. *Analytical and Bioanalytical Chemistry*, *411*, 7187–7196. PMID: 31620825

Zhang, S., Wang, S., & Wu, L. (2010). A novel method for magnetic resonance brain image classification based on adaptive chaotic PSO. *Progress in Electromagnetics Research*, *109*, 325–343.

Zhang, Y. D., Wang, S. H., Ji, G. L., & Dong, Z. C. (2015). Exponential wavelet iterative shrinkage thresholding algorithm with random shift for compressed sensing magnetic resonance imaging. *IEEJ Transactions on Electrical and Electronic Engineering*, *10*, 116–117.

Zhang, Y. D., Wang, S. H., Ji, G. L., & Phillips, P. (2014). Fruit classification using computer vision and feedforward neural network. *Journal of Food Engineering*, *143*, 167–177.

Zhang, Y. D., Wang, S. H., Phillips, P., Dong, Z. C., Ji, G. L., & Yang, J. Q. (2015). Detection of Alzheimer's disease and mild cognitive impairment based on structural volumetric MR images using 3D-DWT and WTA-KSVM trained by PSOTVAC. *Biomedical Signal Processing and Control*, *21*, 58–73.

Zhang, Y., Wang, Y., & Zhang, Y. (2021). Computer Vision Technology in Food Quality Inspection: A Review. *Food Control*, *123*, 107823.

Zhang, Y., Zhou, J. H., Wang, S. M., Wang, Y. Y., Zhang, Y. H., Zhao, S., Liu, S. Y., & Yang, J. (2023). Identification of Xinhui Citri Reticulatae Pericarpium of Different AgingYears Based on Visible Near Infrared Hyperspectral lmaging. *Guangpuxue Yu Guangpu Fenxi*, *43*, 3286–3292.

Zhao, J. W., Chen, Q. S., Huang, X. Y., & Fang, C. H. (2006). Qualitative identification of tea categories by near infrared spectroscopy and support vector machine. *Journal of Pharmaceutical and Biomedical Analysis*, *41*, 1198–1204. PMID: 16621404

About the Contributors

Pethuru Raj, PhD is a chief architect at Reliance Jio Platforms Ltd. (JPL) Bangalore. I previously worked in IBM Global Cloud Centre of Excellence (CoE), Wipro consulting services (WCS), and Robert Bosch Corporate Research (CR). I have gained over 22 years of IT industry experience and eight years of research experience. Finished the CSIR-sponsored PhD degree at Anna University, Chennai, and continued with the UGC-sponsored postdoctoral research in the Department of Computer Science and Automation, Indian Institute of Science (IISc), Bangalore. After that, I was granted two international research fellowships (JSPS and JST) to work as a research scientist for 3.5 years in two leading Japanese universities. I am an IEEE and ACM Professional member. I am focusing on some of the emerging technologies, such as the Internet of Things (IoT), Artificial Intelligence (AI) Model Optimization Techniques, Prompt Engineering, large Language Models (LLMs), Efficient, Explainable, and Edge AI, Blockchain, Digital Twins, Cloud-native and edge computing, Green and Generative AI, High-Performance Computing (HPC) and Quantum Computing, Platform Engineering, 5G/6G, etc.

Kai Sheng received a B.E.(2000) degree in information engineering from Central South University, China, and a PhD.(2006) in risk engineering from the University of Tsukuba, Japan. Now, he is a professor of the AAIR at Xidian University, China. His recent research interests mainly focus on applying AI technologies to V2X-based automated driving(AD), from collaborating environment sensing, communication algorithms, and motion planning to system security and safety.

Kavita Saini is presently working as Professor, School of Computing Science and Engineering, Galgotias University, Delhi NCR, India. She received her PhD degree from Banasthali Vidyapeeth, Banasthali. She has 20 years of teaching and research experience supervising Masters and PhD scholars in emerging technologies. She has published more than 100 research papers in national and international journals and conferences. She has published 20 authored books for UG and PG courses for a number of universities including MD University, Rothak, and Punjab Technical University, Jallandhar with National Publishers. Kavita Saini has edited more than 15 books with International Publishers including Elsevier, IET Publisher, IGI Global and CRC Press, NOVA Science Publisher and published 30 book chapters with International publishers. Under her guidance many M Tech and Ph D scholars are carrying out research work and few have completed under her guidance. Kavita Saini has also delivered technical talks on Blockchain: An Emerging Technology, Web to Deep Web and other emerging Areas and Handled many Special Sessions in International Conferences and Special Issues in International Journals. Her research interests include Smart and Sustainable Agricultural, Blockchain Technology, Industry 5.O, and Cloud and Edge Computing, Digital Twin and Web-Based Instructional Systems (WBIS).

Koteeswaran S, B.Tech., M.E., Ph.D. currently working as Professor and Head in the Department of Computer Science and Engineering (AI and ML), S.A. Engineering College, Chennai-600077, TamilNadu, India. He is having 15 years of teaching experience and published more than 50 research articles in various peer reviewed Journals. He is author for two text books and two edited books for Computer Science & Engineering Programme. His research interests include Artificial Intelligence, Machine Learning, Deep Learning, Big Data and Analytics and Internet of Things. He has presented several papers in conference proceedings. He is a reviewer for more than a dozen journals and also organized more than 25 various events such as National and International Conferences, Faculty Development Programs, Workshops, Seminars, National Level Paper Contests, Quiz programmes, 24 Hours IEEE Xtreme Programming Competition and 36 hours Hachathon. He is a Member of ACM, Member of IAEng, Global Member of ISOC, Member of IEEE CS, Member of

IEEE CIS, Member of IEEE PCS, Member of IET, and a Senior Member of IEEE and also a Life Member of ISTE. He serves as Vice Chairman of ACM Chennai Chapter, Treasurer of IEEE Computer Society & Treasurer cum Secretary of IEEE Professional Communication Society, IEEE Madras Section and also served as Executive Committee Member, Chairman – Paper Contest and Vice Chairman – Student Activities Committee of IEEE Madras Section.

***

A. Muthuram is an Assistant Professor in the Department of Aerospace Engineering at SRM Institute of Science and Technology, Kattankulathur, Tamil Nadu. He is dedicated to teaching and research in the field of aerospace engineering. His areas of interest include aerodynamics, propulsion systems, and advanced aerospace technologies.

Revathi A. (Karthick) is a faculty member in the Department of Computer Science and Engineering (Cyber Security) at R.M.K. College of Engineering and Technology. She is committed to educating students in cybersecurity and advanced computing technologies. Her areas of interest include network security, cryptography, and secure software systems.

A. N. Vinodhini is a faculty member at Erode Sengunthar Engineering College, Thuduppathi, Tamil Nadu. She is actively involved in teaching and academic research in the field of engineering and technology. Her areas of interest include software development, cybersecurity, and emerging trends in computer science.

B. Maheswari is a faculty member from RMK Engineering College, Kavaraipettai, Tamil Nadu. She is involved in teaching and research activities in the field of engineering and technology. Her academic interests include computer science, data analytics, and innovative teaching methodologies.

Maheswari B is an faculty member at R.M.K. Engineering College, Tamil Nadu. She is actively involved in teaching and research, contributing to advancements in engineering and technology. Her work reflects a strong commitment to academic excellence, innovation, and student development.

B. Yamini is a faculty member at SRM Institute of Science and Technology, Kattankulathur. She is actively involved in teaching and research, contributing to academic excellence and innovation. Her areas of interest include machine learning, cybersecurity, and the Internet of Things (IoT).

Hemanth Duraisamy is a 3rd-year BBA General student at Christ University, Delhi NCR, with strong interests in business strategy, data analytics, and finance. Skilled in Python, SQL, Excel, and Power BI, he has applied his knowledge to academic projects on e-commerce insights, marketing strategies, and operations management.

D. Jayasutha is an Associate Professor in the Department of Computer Science and Engineering at Kathir College of Engineering, Coimbatore. She is actively involved in teaching, research, and mentoring students in various domains of computer science. Her areas of expertise include artificial intelligence, machine learning, and data mining.

G. Belshia Jebamalar is an Assistant Professor in the Department of Computer Science and Engineering at S.A. Engineering College, Chennai. She is dedicated to teaching and mentoring students in various areas of computer science. Her academic interests include software engineering, programming languages, and emerging computing technologies.

G. Porkodi is an Assistant Professor in the Department of Computer Science and Business Systems at Vel Tech Multi Tech Dr. Rangarajan Dr. Sankunthala Engineering College, Avadi, Chennai. She is dedicated to teaching and guiding students in the intersection of technology and business systems. Her academic interests include data analytics, business intelligence, and software development.

Monica G. K. is a faculty member in the Department of Computer Science and Engineering (Cyber Security) at R.M.K. College of Engineering and Technology. She is involved in teaching and mentoring students in the areas of cybersecurity and advanced computing. Her interests include network security, ethical hacking, and secure software development.

K. Bhavani is an Assistant Professor in the Department of Computer Science and Business Systems at Panimalar Engineering College, Chennai. She is dedicated to teaching and shaping young minds in the field of computer science with a business perspective. Her academic interests include data analytics, business systems, and modern software technologies.

R. Kennady is a faculty member in the Department of Artificial Intelligence and Data Science at Rajalakshmi Institute of Technology, Chennai. His areas of interest include artificial intelligence, data analytics, and intelligent systems.

Nivas Kumar is currently affiliated with Jawahar Engineering College, Saligramam, Chennai, where he contributes to academic and research activities in the field of engineering. He holds expertise in Mechanical Engineering and is actively involved in student mentorship and departmental initiatives. His professional interests include advancements in thermal systems, design engineering, and sustainable technologies.

M. Anita is a faculty member at SRM Institute of Science and Technology, Kattankulathur, Tamil Nadu. She is actively engaged in teaching, research, and academic development in the field of engineering and technology. Her areas of expertise include artificial intelligence, machine learning, and data science.

Nalini M is an faculty member in the Department of Cyber Security at S.A. Engineering College, Chennai. Her research interests include machine learning, deep learning, and their applications in cybersecurity

M. Sukanya is an Assistant Professor at Kathir College of Engineering, Coimbatore. She is actively engaged in teaching and research in the field of engineering and technology. Her academic interests include data science, machine learning, and software engineering.

G. Dinesh is a faculty member at EASA College of Engineering and Technology, Coimbatore, Tamil Nadu. He is actively involved in teaching and research, with a strong focus on engineering education and technological advancements. His areas of interest include machine learning, data analytics, and smart systems development.

M. Gowri is an Assistant Professor at Sathyabama Institute of Science and Technology, Chennai. She is actively involved in teaching and academic research in the field of engineering and technology. Her areas of interest include computer science, data analytics, and emerging technologies.

N. Pavithra is an Assistant Professor in the Department of Information Technology at Jeppiaar Institute of Technology. She is committed to academic excellence and guiding students in the field of computing technologies. Her areas of interest include software engineering, cybersecurity, and emerging IT trends.

N. Subhash Chandra is a Professor at CVR College of Engineering, Hyderabad, Telangana. He is committed to academic excellence and research in the field of engineering and technology. His areas of expertise include computer networks, cybersecurity, and advanced computing systems.

Aparna P is a faculty member in the Department of Computer Science and Engineering at RMK College of Engineering and Technology, RSM Nagar, Puduvoyal – 601206. She is committed to academic excellence and student development in the field of computer science. Her areas of interest include programming, data structures, and emerging computing technologies.

P. Girija is a faculty member in the Department of Information Technology at Panimalar Engineering College, Chennai. She is actively involved in teaching and research at Panimalar Engineering College, focusing on student development and technical excellence. Her areas of specialization include data science, computer networks, and artificial intelligence.

Vivek Pandiya Raj is an Assistant Professor (Senior Grade) in the Department of Computer Science and Engineering at the School of Computing, Vel Tech Dr. Sagunthala R&D Institute of Science and Technology, Chennai. He specializes in computer science education and research, contributing to both academic instruction and scholarly publications. With a strong focus on innovation and student development, he actively engages in research-driven teaching and technical mentoring

Shrikant Patel is working as an Associate Professor in the Department of Computer Science and Applications at Delhi Skill and Entrepreneurship University, Delhi (Govt. of NCT of Delhi), India. He has more than 20 years of teaching experience in various leading institutes/organizations at both UG & PG levels. He has published about more than 25 international research papers, filed 5 patents, 3 books and guided more than 250 students for their minor and major research projects in various subfields of Computer Science and Information Technology.

Siva Subramanian R is a faculty member at SRM Institute of Science and Technology, Trichy campus. As an active academic and researcher, he is involved in advancing knowledge in the field of engineering and technology. His areas of interest include artificial intelligence, machine learning, and data-driven solutions.

Tamilvanddan R has a Bachelor of technology in information technology at Hindustan institute of technology and science

S. Alfiya is an Assistant Professor in the Department of Computer Science and Engineering at Vel Tech Multi Tech Dr. Rangarajan Dr. Sakunthala Engineering College, Avadi, Chennai. She is actively involved in teaching and mentoring students in core computer science subjects. Her academic interests include emerging technologies and their applications in software development and engineering.

Palak Saini is pursuing a Bachelor of Technology (B.Tech) degree in Biotechnology at Sharda University, India. Her undergraduate research experience includes projects on Iron and Iron Oxide Nanoparticles and the Assessment of Microbial Diversity for Pigment Production. She is currently working on Silver Nanoparticles and Their Antimicrobial Properties. Her academic interests focus on nanobiotechnology, microbial biotechnology, and their applications in health and industry.

C.V. Suresh Babu is a pioneer in content development. A true entrepreneur, he founded Anniyappa Publications, a company that is highly active in publishing books related to Computer Science and Management. Dr. C.V. Suresh Babu has also ventured into SB Institute, a center for knowledge transfer. He holds a Ph.D. in Engineering Education from the National Institute of Technical Teachers Training & Research in Chennai, along with seven master's degrees in various disciplines such as Engineering, Computer Applications, Management, Commerce, Economics, Psychology, Law, and Education. Additionally, he has UGC-NET/SET qualifications in the fields of Computer Science, Management, Commerce, and Education. Currently, Dr. C.V. Suresh Babu is a Professor in the Department of Information Technology at the School of Computing Science, Hindustan Institute of Technology and Science (Hindustan University) in Padur, Chennai, Tamil Nadu, India. For more information, you can visit his personal blog at https://sites.google.com/view/cvsureshbabu/.

Thilagam T is an accomplished academic and researcher in the Department of Computer Science and Engineering at Vel Tech Rangarajan Dr. Sagunthala R&D Institute of Science and Technology, Chennai. Her expertise spans machine learning, artificial intelligence, and data analytics. She is dedicated to advancing research and innovation in emerging computing technologies.

Anish T. P is an Assistant Professor in the Department of Computer Science and Engineering at R.M.K. College of Engineering and Technology, Puduvoyal, Tamil Nadu. He is dedicated to teaching, research, and student mentorship in core and advanced areas of computer science. His interests include artificial intelligence, cybersecurity, and modern software development practices.

Index

Symbols

A

C

D

E

F

G

I

L

M

O

P

Q

S

T

www.ingramcontent.com/pod-product-compliance
Lightning Source LLC
LaVergne TN
LVHW081524300926
849161LV00006B/1290